RAG-Driven Generative

Build custom retrieval augmented generation pipelines with
LlamaIndex, Deep Lake, and Pinecone

Denis Rothman

RAG-Driven Generative AI

Copyright © 2024 Packt Publishing

Senior Publishing Product Manager: Bhavesh Amin

Acquisition Editor – Peer Reviews: Swaroop Singh

Project Editor: Janice Gonsalves

Content Development Editor: Tanya D'cruz

Copy Editor: Safis Editor

Technical Editor: Karan Sonawane

Proofreader: Safis Editor

Indexer: Rekha Nair

Presentation Designer: Ajay Patule

Developer Relations Marketing Executive: Anamika Singh

First published: September 2024

Production reference: 3181225

Published by Packt Publishing Ltd.
Livery Place
35 Livery Street
Birmingham
B3 2PB, UK.

ISBN: 978-1-83620-091-8

www.packt.com

Contributors

About the author

Denis Rothman graduated from Sorbonne University and Paris-Diderot University, and as a student, he wrote and registered a patent for one of the earliest word2vector embeddings and word piece tokenization solutions. He started a company focused on deploying AI and went on to author one of the first AI cognitive NLP chatbots, applied as a language teaching tool for Moët et Chandon (part of LVMH) and more. Denis rapidly became an expert in explainable AI, incorporating interpretable, acceptance-based explanation data and interfaces into solutions implemented for major corporate projects in the aerospace, apparel, and supply chain sectors. His core belief is that you only really know something once you have taught somebody how to do it.

About the reviewers

Alberto Romero has always had a passion for technology and open source, from programming at the age of 12 to hacking the Linux kernel by 14 back in the 90s. In 2017, he co-founded an AI startup and served as its CTO for six years, building an award-winning InsurTech platform from scratch. He currently continues to design and build generative AI platforms in financial services, leading multiple initiatives in this space. He has developed and productionized numerous AI products that automate and improve decision-making processes, already serving thousands of users. He serves as an advisor to an advanced data security and governance startup that leverages predictive ML and Generative AI to address modern enterprise data security challenges.

I would like to express my deepest gratitude to my wife, Alicia, and daughters, Adriana and Catalina, for their unwavering support throughout the process of reviewing this book. Their patience, encouragement, and love have been invaluable, and I am truly fortunate to have them by my side.

Shubham Garg is a senior applied scientist at Amazon, specializing in developing **Large Language Models (LLMs)** and **Vision-Language Models (VLMs)**. He has led innovative projects at Amazon and IBM, including developing Alexa's translation features, dynamic prompt construction, and optimizing AI tools. Shubham has contributed to advancements in NLP, multilingual models, and AI-driven solutions. He has published at major NLP conferences, reviewed for conferences and journals, and holds a patent. His deep expertise in AI technologies makes his perspective as a reviewer both valuable and insightful.

Tamilselvan Subramanian is a seasoned AI leader and two-time founder, specializing in generative AI for text and images. He has built and scaled AI-driven products, including an AI conservation platform to save endangered species, a medical image diagnostic platform, an AI-driven EV leasing platform, and an Enterprise AI platform from scratch. Tamil has authored multiple AI articles published in medical journals and holds two patents in AI and image processing. He has served as a technical architect and consultant for finance and energy companies across Europe, the US, and Australia, and has also worked for IBM and Wipro. Currently, he focuses on cutting-edge applications of computer vision, text, and generative AI.

My special thanks go to my wife Suganthi, my son Sanjeev, and my mom and dad for their unwavering support, allowing me the personal time to work on this book.

Learn further in a live workshop with the author

If you are curious how RAG-based systems scale into agentic workflows using context engineering, consider registering for the author's live workshop on 24 January 2026, which extends the ideas discussed here into hands-on practice.

https://packt.link/gk5Yu

Table of Contents

Chapter 2: RAG Embedding Vector Stores with Deep Lake and OpenAI   31

Chapter 9: Empowering AI Models: Fine-Tuning RAG Data and Human Feedback 235

Chapter 10: RAG for Video Stock Production with Pinecone and OpenAI 255

Appendix   293

Other Books You May Enjoy   305

Index   309

Preface

Designing and managing controlled, reliable, multimodal generative AI pipelines is complex. *RAG-Driven Generative AI* provides a roadmap for building effective LLM, computer vision, and generative AI systems that will balance performance and costs.

From foundational concepts to complex implementations, this book offers a detailed exploration of how RAG can control and enhance AI systems by tracing each output to its source document. RAG's traceable process allows human feedback for continual improvements, minimizing inaccuracies, hallucinations, and bias. This AI book shows you how to build a RAG framework from scratch, providing practical knowledge on vector stores, chunking, indexing, and ranking. You'll discover techniques in optimizing performance and costs, improving model accuracy by integrating human feedback, balancing costs with when to fine-tune, and improving accuracy and retrieval speed by utilizing embedded-indexed knowledge graphs.

Experience a blend of theory and practice using frameworks like LlamaIndex, Pinecone, and Deep Lake and generative AI platforms such as OpenAI and Hugging Face.

By the end of this book, you will have acquired the skills to implement intelligent solutions, keeping you competitive in fields from production to customer service across any project.

Who this book is for

This book is ideal for data scientists, AI engineers, machine learning engineers, and MLOps engineers, as well as solution architects, software developers, and product and project managers working on LLM and computer vision projects who want to learn and apply RAG for real-world applications. Researchers and natural language processing practitioners working with large language models and text generation will also find the book useful.

What this book covers

Chapter 1, Why Retrieval Augmented Generation?, introduces RAG's foundational concepts, outlines its adaptability across different data types, and navigates the complexities of integrating the RAG framework into existing AI platforms. By the end of this chapter, you will have gained a solid understanding of RAG and practical experience in building diverse RAG configurations for naïve, advanced, and modular RAG using Python, preparing you for more advanced applications in subsequent chapters.

Chapter 2, RAG Embedding Vector Stores with Deep Lake and OpenAI, dives into the complexities of RAG-driven generative AI by focusing on embedding vectors and their storage solutions. We explore the transition from raw data to organized vector stores using Activeloop Deep Lake and OpenAI models, detailing the process of creating and managing embeddings that capture deep semantic meanings. You will learn to build a scalable, multi-team RAG pipeline from scratch in Python by dissecting the RAG ecosystem into independent components. By the end, you'll be equipped to handle large datasets with sophisticated retrieval capabilities, enhancing generative AI outputs with embedded document vectors.

Chapter 3, Building Index-Based RAG with LlamaIndex, Deep Lake, and OpenAI, dives into index-based RAG, focusing on enhancing AI's precision, speed, and transparency through indexing. We'll see how LlamaIndex, Deep Lake, and OpenAI can be integrated to put together a traceable and efficient RAG pipeline. Through practical examples, including a domain-specific drone technology project, you will learn to manage and optimize index-based retrieval systems. By the end, you will be proficient in using various indexing types and understand how to enhance the data integrity and quality of your AI outputs.

Chapter 4, Multimodal Modular RAG for Drone Technology, raises the bar of all generative AI applications by introducing a multimodal modular RAG framework tailored for drone technology. We'll develop a generative AI system that not only processes textual information but also integrates advanced image recognition capabilities. You'll learn to build and optimize a Python-based multimodal modular RAG system, using tools like LlamaIndex, Deep Lake, and OpenAI, to produce rich, context-aware responses to queries.

Chapter 5, Boosting RAG Performance with Expert Human Feedback, introduces adaptive RAG, an innovative enhancement to standard RAG that incorporates human feedback into the generative AI process. By integrating expert feedback directly, we will create a hybrid adaptive RAG system using Python, exploring the integration of human feedback loops to refine data continuously and improve the relevance and accuracy of AI responses.

Chapter 6, Scaling RAG Bank Customer Data with Pinecone, guides you through building a recommendation system to minimize bank customer churn, starting with data acquisition and exploratory analysis using a Kaggle dataset. You'll move onto embedding and upserting large data volumes with Pinecone and OpenAI's technologies, culminating in developing AI-driven recommendations with GPT-4o. By the end, you'll know how to implement advanced vector storage techniques and AI-driven analytics to enhance customer retention strategies.

Chapter 7, Building Scalable Knowledge-Graph-Based RAG with Wikipedia API and LlamaIndex, details the development of three pipelines: data collection from Wikipedia, populating a Deep Lake vector store, and implementing a knowledge graph index-based RAG. You'll learn to automate data retrieval and preparation, create and query a knowledge graph to visualize complex data relationships, and enhance AI-generated responses with structured data insights. You'll be equipped by the end to build and manage a knowledge graph-based RAG system, providing precise, context-aware output.

Chapter 8, Dynamic RAG with Chroma and Hugging Face Llama, explores dynamic RAG using Chroma and Hugging Face's Llama technology. It introduces the concept of creating temporary data collections daily, optimized for specific meetings or tasks, which avoids long-term data storage issues. You will learn to build a Python program that manages and queries these transient datasets efficiently, ensuring that the most relevant and up-to-date information supports every meeting or decision point. By the end, you will be able to implement dynamic RAG systems that enhance responsiveness and precision in data-driven environments.

Chapter 9, Empowering AI Models: Fine-Tuning RAG Data and Human Feedback, focuses on fine-tuning techniques to streamline RAG data, emphasizing how to transform extensive, non-parametric raw data into a more manageable, parametric format with trained weights suitable for continued AI interactions. You'll explore the process of preparing and fine-tuning a dataset, using OpenAI's tools to convert data into prompt and completion pairs for machine learning. Additionally, this chapter will guide you through using OpenAI's GPT-4o-mini model for fine-tuning, assessing its efficiency and cost-effectiveness.

Chapter 10, RAG for Video Stock Production with Pinecone and OpenAI, explores the integration of RAG in video stock production, combining human creativity with AI-driven automation. It details constructing an AI system that produces, comments on, and labels video content, using OpenAI's text-to-video and vision models alongside Pinecone's vector storage capabilities. Starting with video generation and technical commentary, the journey extends to managing embedded video data within a Pinecone vector store.

To get the most out of this book

You should have basic **Natural Processing Language** (NLP) knowledge and some experience with Python. Additionally, most of the programs in this book are provided as Jupyter notebooks. To run them, all you need is a free Google Gmail account, allowing you to execute the notebooks on Google Colaboratory's free **virtual machine** (VM). You will also need to generate API tokens for OpenAI, Activeloop, and Pinecone.

The following modules will need to be installed when running the notebooks:

Modules	Version
deeplake	3.9.18 (with Pillow)
openai	1.40.3 (requires regular upgrades)
transformers	4.41.2
numpy	>=1.24.1 (Upgraded to satisfy chex)
deepspeed	0.10.1
bitsandbytes	0.41.1
accelerate	0.31.0
tqdm	4.66.1
neural_compressor	2.2.1
onnx	1.14.1

pandas	2.0.3
scipy	1.11.2
beautifulsoup4	4.12.3
requests	2.31.0

Download the example code files

The code bundle for the book is hosted on GitHub at https://github.com/Denis2054/RAG-Driven-Generative-AI. We also have other code bundles from our rich catalog of books and videos available at https://github.com/PacktPublishing/. Check them out!

Download the color images

We also provide a PDF file that has color images of the screenshots/diagrams used in this book. You can download it here: https://packt.link/gbp/9781836200918.

Conventions used

There are a number of text conventions used throughout this book.

CodeInText: Indicates code words in text, database table names, folder names, filenames, file extensions, pathnames, dummy URLs, user input, and Twitter handles. For example: "self refers to the current instance of the class to access its variables, methods, and functions".

A block of code is set as follows:

```
# Cosine Similarity
score = calculate_cosine_similarity(query, best_matching_record)
print(f"Best Cosine Similarity Score: {score:.3f}")
```

Any command-line input or output is written as follows:

```
Best Cosine Similarity Score: 0.126
```

Bold: Indicates a new term, an important word, or words that you see on the screen. For example, text in menus or dialog boxes appears like this. Here is an example: "**Modular RAG** implementing flexible retrieval methods".

 Warnings or important notes appear like this.

 Tips and tricks appear like this.

Get in touch

Feedback from our readers is always welcome.

General feedback: Email feedback@packtpub.com, and mention the book's title in the subject of your message. If you have questions about any aspect of this book, please email us at questions@packtpub.com.

Errata: Although we have taken every care to ensure the accuracy of our content, mistakes do happen. If you have found a mistake in this book we would be grateful if you would report this to us. Please visit http://www.packtpub.com/submit-errata, select your book, click on the Errata Submission Form link, and enter the details.

Piracy: If you come across any illegal copies of our works in any form on the Internet, we would be grateful if you would provide us with the location address or website name. Please contact us at copyright@packtpub.com with a link to the material.

If you are interested in becoming an author: If there is a topic that you have expertise in and you are interested in either writing or contributing to a book, please visit http://authors.packtpub.com.

Share your thoughts

Once you've read *RAG-Driven Generative AI*, we'd love to hear your thoughts! Scan the QR code below to go straight to the Amazon review page for this book and share your feedback.

https://packt.link/r/1836200919

Your review is important to us and the tech community and will help us make sure we're delivering excellent quality content.

Making the Most Out of This Book — Get to Know Your Free Benefits

Unlock exclusive free benefits that come with your purchase, thoughtfully crafted to super-charge your learning journey and help you learn without limits.

UNLOCK NOW

Note: Have your purchase invoice ready before you begin. https://www.packtpub.com/unlock/9781836200918

Figure 1.1: Next-Gen Reader, AI Assistant (Beta), and Free PDF access

Enhanced reading experience with our Next-gen Reader:

    ↻ Multi-device progress sync: Learn from any device with seamless progress sync.

    📑 Highlighting and Notetaking: Turn your reading into lasting knowledge.

    🔖 Bookmarking: Revisit your most important learnings anytime.

    ☀ Dark mode: Focus with minimal eye strain by switching to dark or sepia modes.

Learn smarter using our AI assistant (Beta):

    ✦ Summarize it: Summarize key sections or an entire chapter.

    ✦ AI code explainers: In Packt Reader, click the "Explain" button above each code block for AI-powered code explanations.

Note: AI Assistant is part of next-gen Packt Reader and is still in beta.

Learn anytime, anywhere:

    📄📘 Access your content offline with DRM-free PDF and ePub versions—compatible with your favorite e-readers.

Unlock Your Book's Exclusive Benefits

Your copy of this book comes with the following exclusive benefits:

- ☁ Next-gen Packt Reader
- ✦ AI assistant (beta)
- 📄 DRM-free PDF/ePub downloads

Use the following guide to unlock them if you haven't already. The process takes just a few minutes and needs to be done only once.

How to unlock these benefits in three easy steps

Step 1

Have your purchase invoice for this book ready, as you'll need it in *Step 3*. If you received a physical invoice, scan it on your phone and have it ready as either a PDF, JPG, or PNG.

For more help on finding your invoice, visit `https://www.packtpub.com/unlock-benefits/help`.

 Note: Bought this book directly from Packt? You don't need an invoice. After completing Step 2, you can jump straight to your exclusive content.

Step 2

Scan the following QR code or visit `https://www.packtpub.com/unlock/9781836200918`:

Step 3

Sign in to your Packt account or create a new one for free. Once you're logged in, upload your invoice. It can be in PDF, PNG, or JPG format and must be no larger than 10 MB. Follow the rest of the instructions on the screen to complete the process.

Need help?

If you get stuck and need help, visit `https://www.packtpub.com/unlock-benefits/help` for a detailed FAQ on how to find your invoices and more. The following QR code will take you to the help page directly:

 Note: If you are still facing issues, reach out to `customercare@packt.com`.

1

Why Retrieval Augmented Generation?

Even the most advanced generative AI models can only generate responses based on the data they have been trained on. They cannot provide accurate answers to questions about information outside their training data. Generative AI models simply don't know that they don't know! This leads to inaccurate or inappropriate outputs, sometimes called hallucinations, bias, or, simply said, nonsense.

Retrieval Augmented Generation (RAG) is a framework that addresses this limitation by combining retrieval-based approaches with generative models. It retrieves relevant data from external sources in real time and uses this data to generate more accurate and contextually relevant responses. Generative AI models integrated with RAG retrievers are revolutionizing the field with their unprecedented efficiency and power. One of the key strengths of RAG is its adaptability. It can be seamlessly applied to any type of data, be it text, images, or audio. This versatility makes RAG ecosystems a reliable and efficient tool for enhancing generative AI capabilities.

A project manager, however, already encounters a wide range of generative AI platforms, frameworks, and models such as Hugging Face, Google Vertex AI, OpenAI, LangChain, and more. An additional layer of emerging RAG frameworks and platforms will only add complexity with Pinecone, Chroma, Activeloop, LlamaIndex, and so on. All these Generative AI and RAG frameworks often overlap, creating an incredible number of possible configurations. Finding the right configuration of models and RAG resources for a specific project, therefore, can be challenging for a project manager. There is no silver bullet. The challenge is tremendous, but the rewards, when achieved, are immense!

We will begin this chapter by defining the RAG framework at a high level. Then, we will define the three main RAG configurations: naïve RAG, advanced RAG, and modular RAG. We will also compare RAG and fine-tuning and determine when to use these approaches. RAG can only exist within an ecosystem, and we will design and describe one in this chapter. Data needs to come from somewhere and be processed. Retrieval requires an organized environment to retrieve data, and generative AI models have input constraints.

Finally, we will dive into the practical aspect of this chapter. We will build a Python program from scratch to run entry-level naïve RAG with keyword search and matching. We will also code an advanced RAG system with vector search and index-based retrieval. Finally, we will build a modular RAG that takes both naïve and advanced RAG into account. By the end of this chapter, you will acquire a theoretical understanding of the RAG framework and practical experience in building a RAG-driven generative AI program. This hands-on approach will deepen your understanding and equip you for the following chapters.

In a nutshell, this chapter covers the following topics:

- Defining the RAG framework
- The RAG ecosystem
- Naïve keyword search and match RAG in Python
- Advanced RAG with vector-search and index-based RAG in Python
- Building a modular RAG program

Let's begin by defining RAG.

What is RAG?

When a generative AI model doesn't know how to answer accurately, some say it is hallucinating or producing bias. Simply said, it just produces nonsense. However, it all boils down to the impossibility of providing an adequate response when the model's training didn't include the information requested beyond the classical model configuration issues. This confusion often leads to random sequences of the most probable outputs, not the most accurate ones.

RAG begins where generative AI ends by providing the information an LLM model lacks to answer accurately. RAG was designed (Lewis et al., 2020) for LLMs. The RAG framework will perform optimized information retrieval tasks, and the generation ecosystem will add this information to the input (user query or automated prompt) to produce improved output. The RAG framework can be summed up at a high level in the following figure:

Figure 1.1: The two main components of RAG-driven generative AI

Think of yourself as a student in a library. You have an essay to write on RAG. Like ChatGPT, for example, or any other AI copilot, you have learned how to read and write. As with any **Large Language Model (LLM)**, you are sufficiently trained to read advanced information, summarize it, and write content. However, like any superhuman AI you will find from Hugging Face, Vertex AI, or OpenAI, there are many things you don't know.

In the *retrieval* phase, you search the library for books on the topic you need (the left side of *Figure 1.1*). Then, you go back to your seat, perform a retrieval task by yourself or a co-student, and extract the information you need from those books. In the *generation* phase (the right side of *Figure 1.1*), you begin to write your essay. You are a RAG-driven generative human agent, much like a RAG-driven generative AI framework.

As you continue to write your essay on RAG, you stumble across some tough topics. You don't have the time to go through all the information available physically! You, as a generative human agent, are stuck, just as a generative AI model would be. You may try to write something, just as a generative AI model does when its output makes little sense. But you, like the generative AI agent, will not realize whether the content is accurate or not until somebody corrects your essay and you get a grade that will rank your essay.

At this point, you have reached your limit and decide to turn to a RAG generative AI copilot to ensure you get the correct answers. However, you are puzzled by the number of LLM models and RAG configurations available. You need first to understand the resources available and how RAG is organized. Let's go through the main RAG configurations.

Naïve, advanced, and modular RAG configurations

A RAG framework necessarily contains two main components: a retriever and a generator. The generator can be any LLM or foundation multimodal AI platform or model, such as GPT-4o, Gemini, Llama, or one of the hundreds of variations of the initial architectures. The retriever can be any of the emerging frameworks, methods, and tools such as Activeloop, Pinecone, LlamaIndex, LangChain, Chroma, and many more.

The issue now is to decide which of the three types of RAG frameworks (Gao et al., 2024) will fit the needs of a project. We will illustrate these three approaches in code in the *Naïve, advanced, and modular RAG in code* section of this chapter:

- **Naïve RAG**: This type of RAG framework doesn't involve complex data embedding and indexing. It can be efficient to access reasonable amounts of data through keywords, for example, to augment a user's input and obtain a satisfactory response.

- **Advanced RAG**: This type of RAG involves more complex scenarios, such as with vector search and indexed-base retrieval applied. Advanced RAG can be implemented with a wide range of methods. It can process multiple data types, as well as multimodal data, which can be structured or unstructured.

- **Modular RAG**: Modular RAG broadens the horizon to include any scenario that involves naïve RAG, advanced RAG, machine learning, and any algorithm needed to complete a complex project.

However, before going further, we need to decide if we should implement RAG or fine-tune a model.

RAG versus fine-tuning

RAG is not always an alternative to fine-tuning, and fine-tuning cannot always replace RAG. If we accumulate too much data in RAG datasets, the system may become too cumbersome to manage. On the other hand, we cannot fine-tune a model with dynamic, ever-changing data such as daily weather forecasts, stock market values, corporate news, and all forms of daily events.

The decision of whether to implement RAG or fine-tune a model relies on the proportion of parametric versus non-parametric information. The fundamental difference between a model trained from scratch or fine-tuned and RAG can be summed up in terms of parametric and non-parametric knowledge:

- **Parametric**: In a RAG-driven generative AI ecosystem, the parametric part refers to the generative AI model's parameters (weights) learned through training data. This means the model's knowledge is stored in these learned weights and biases. The original training data is transformed into a mathematical form, which we call a parametric representation. Essentially, the model "remembers" what it learned from the data, but the data itself is not stored explicitly.

- **Non-Parametric**: In contrast, the non-parametric part of a RAG ecosystem involves storing explicit data that can be accessed directly. This means that the data remains available and can be queried whenever needed. Unlike parametric models, where knowledge is embedded indirectly in the weights, non-parametric data in RAG allows us to see and use the actual data for each output.

The difference between RAG and fine-tuning relies on the amount of static (parametric) and dynamic (non-parametric) ever-evolving data the generative AI model must process. A system that relies too heavily on RAG might become overloaded and cumbersome to manage. A system that relies too much on fine-tuning a generative model will display its inability to adapt to daily information updates.

There is a decision-making threshold illustrated in *Figure 1.2* that shows that a RAG-driven generative AI project manager will have to evaluate the potential of the ecosystem's trained parametric generative AI model before implementing a non-parametric (explicit data) RAG framework. The potential of the RAG component requires careful evaluation as well.

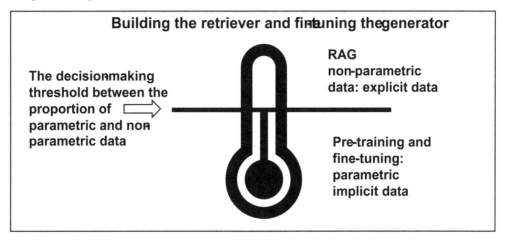

Figure 1.2: The decision-making threshold between enhancing RAG or fine-tuning an LLM

In the end, the balance between enhancing the retriever and the generator in a RAG-driven generative AI ecosystem depends on a project's specific requirements and goals. RAG and fine-tuning are not mutually exclusive.

RAG can be used to improve a model's overall efficiency, together with fine-tuning, which serves as a method to enhance the performance of both the retrieval and generation components within the RAG framework. We will fine-tune a proportion of the retrieval data in *Chapter 9, Empowering AI Models: Fine-Tuning RAG Data and Human Feedback*.

We will now see how a RAG-driven generative AI involves an ecosystem with many components.

The RAG ecosystem

RAG-driven generative AI is a framework that can be implemented in many configurations. RAG's framework runs within a broad ecosystem, as shown in *Figure 1.3*. However, no matter how many retrieval and generation frameworks you encounter, it all boils down to the following four domains and questions that go with them:

- **Data:** Where is the data coming from? Is it reliable? Is it sufficient? Are there copyright, privacy, and security issues?

- **Storage:** How is the data going to be stored before or after processing it? What amount of data will be stored?
- **Retrieval:** How will the correct data be retrieved to augment the user's input before it is sufficient for the generative model? What type of RAG framework will be successful for a project?
- **Generation:** Which generative AI model will fit into the type of RAG framework chosen?

The data, storage, and generation domains depend heavily on the type of RAG framework you choose. Before making that choice, we need to evaluate the proportion of parametric and non-parametric knowledge in the ecosystem we are implementing. *Figure 1.3* represents the RAG framework, which includes the main components regardless of the types of RAG implemented:

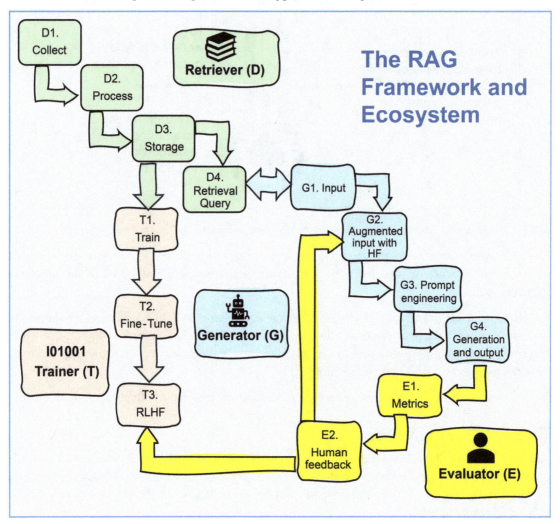

Figure 1.3: The Generative RAG-ecosystem

- The **Retriever (D)** handles data collection, processing, storage, and retrieval
- The **Generator (G)** handles input augmentation, prompt engineering, and generation

- The **Evaluator** (E) handles mathematical metrics, human evaluation, and feedback
- The **Trainer** (T) handles the initial pre-trained model and fine-tuning the model

Each of these four components relies on their respective ecosystems, which form the overall RAG-driven generative AI pipeline. We will refer to the domains D, G, E, and T in the following sections. Let's begin with the retriever.

The retriever (D)

The retriever component of a RAG ecosystem collects, processes, stores, and retrieves data. The starting point of a RAG ecosystem is thus an ingestion data process, of which the first step is to collect data.

Collect (D1)

In today's world, AI data is as diverse as our media playlists. It can be anything from a chunk of text in a blog post to a meme or even the latest hit song streamed through headphones. And it doesn't stop there—the files themselves come in all shapes and sizes. Think of PDFs filled with all kinds of details, web pages, plain text files that get straight to the point, neatly organized JSON files, catchy MP3 tunes, videos in MP4 format, or images in PNG and JPG.

Furthermore, a large proportion of this data is unstructured and found in unpredictable and complex ways. Fortunately, many platforms, such as Pinecone, OpenAI, Chroma, and Activeloop, provide ready-to-use tools to process and store this jungle of data.

Process (D2)

In the data collection phase (D1) of multimodal data processing, various types of data, such as text, images, and videos, can be extracted from websites using web scraping techniques or any other source of information. These data objects are then transformed to create uniform feature representations. For example, data can be chunked (broken into smaller parts), embedded (transformed into vectors), and indexed to enhance searchability and retrieval efficiency.

We will introduce these techniques, starting with the *Building Hybrid Adaptive RAG in Python* section of this chapter. In the following chapters, we will continue building more complex data processing functions.

Storage (D3)

At this stage of the pipeline, we have collected and begun processing a large amount of diverse data from the internet—videos, pictures, texts, you name it. Now, what can we do with all that data to make it useful?

That's where vector stores like Deep Lake, Pinecone, and Chroma come into play. Think of these as super smart libraries that don't just store your data but convert it into mathematical entities as vectors, enabling powerful computations. They can also apply a variety of indexing methods and other techniques for rapid access.

Instead of keeping the data in static spreadsheets and files, we turn it into a dynamic, searchable system that can power anything from chatbots to search engines.

Retrieval query (D4)

The retrieval process is triggered by the user input or automated input (G1).

To retrieve data quickly, we load it into vector stores and datasets after transforming it into a suitable format. Then, using a combination of keyword searches, smart embeddings, and indexing, we can retrieve the data efficiently. Cosine similarity, for example, finds items that are closely related, ensuring that the search results are not just fast but also highly relevant.

Once the data is retrieved, we then augment the input.

The generator (G)

The lines are blurred in the RAG ecosystem between input and retrieval, as shown in *Figure 1.3*, representing the RAG framework and ecosystem. The user input (G1), automated or human, interacts with the retrieval query (D4) to augment the input before sending it to the generative model.

The generative flow begins with an input.

Input (G1)

The input can be a batch of automated tasks (processing emails, for example) or human prompts through a **User Interface** (UI). This flexibility allows you to seamlessly integrate AI into various professional environments, enhancing productivity across industries.

Augmented input with HF (G2)

Human feedback (HF) can be added to the input, as described in the *Human feedback (E2) under Evaluator (E)* section. Human feedback will make a RAG ecosystem considerably adaptable and provide full control over data retrieval and generative AI inputs. In the *Building hybrid adaptive RAG in Python* section of this chapter, we will build augmented input with human feedback.

Prompt engineering (G3)

Both the retriever (D) and the generator (G) rely heavily on prompt engineering to prepare the standard and augmented message that the generative AI model will have to process. Prompt engineering brings the retriever's output and the user input together.

Generation and output (G4)

The choice of a generative AI model depends on the goals of a project. Llama, Gemini, GPT, and other models can fit various requirements. However, the prompt must meet each model's specifications. Frameworks such as LangChain, which we will implement in this book, help streamline the integration of various AI models into applications by providing adaptable interfaces and tools.

The evaluator (E)

We often rely on mathematical metrics to assess the performance of a generative AI model. However, these metrics only give us part of the picture. It's important to remember that the ultimate test of an AI's effectiveness comes down to human evaluation.

Metrics (E1)

A model cannot be evaluated without mathematical metrics, such as cosine similarity, as with any AI system. These metrics ensure that the retrieved data is relevant and accurate. By quantifying the relationships and relevance of data points, they provide a solid foundation for assessing the model's performance and reliability.

Human feedback (E2)

No generative AI system, whether RAG-driven or not, and whether the mathematical metrics seem sufficient or not, can elude human evaluation. It is ultimately human evaluation that decides if a system designed for human users will be accepted or rejected, praised or criticized.

Adaptive RAG introduces the human, real-life, pragmatic feedback factor that will improve a RAG-driven generative AI ecosystem. We will implement adaptive RAG in *Chapter 5*, *Boosting RAG Performance with Expert Human Feedback*.

The trainer (T)

A standard generative AI model is pre-trained with a vast amount of general-purpose data. Then, we can fine-tune (T2) the model with domain-specific data.

We will take this further by integrating static RAG data into the fine-tuning process in *Chapter 9*, *Empowering AI Models: Fine-Tuning RAG Data and Human Feedback*. We will also integrate human feedback, which provides valuable information that can be integrated into the fine-tuning process in a variant of **Reinforcement Learning from Human Feedback** (RLHF).

We are now ready to code entry-level naïve, advanced, and modular RAG in Python.

Naïve, advanced, and modular RAG in code

This section introduces naïve, advanced, and modular RAG through basic educational examples. The program builds keyword matching, vector search, and index-based retrieval methods. Using OpenAI's GPT models, it generates responses based on input queries and retrieved documents.

The goal of the notebook is for a conversational agent to answer questions on RAG in general. We will build the retriever from the bottom up, from scratch, in Python and run the generator with OpenAI GPT-4o in eight sections of code divided into two parts:

Part 1: Foundations and Basic Implementation

1. **Environment** setup for OpenAI API integration
2. **Generator** function using GPT-4o
3. **Data** setup with a list of documents (db_records)
4. **Query** for user input

Part 2: Advanced Techniques and Evaluation

1. **Retrieval metrics** to measure retrieval responses

2. **Naïve RAG** with a keyword search and matching function

3. **Advanced RAG** with vector search and index-based search

4. **Modular RAG** implementing flexible retrieval methods

To get started, open `RAG_Overview.ipynb` in the GitHub repository. We will begin by establishing the foundations of the notebook and exploring the basic implementation.

Part 1: Foundations and basic implementation

In this section, we will set up the environment, create a function for the generator, define a function to print a formatted response, and define the user query.

The first step is to install the environment.

 The section titles of the following implementation of the notebook follow the structure in the code. Thus, you can follow the code in the notebook or read this self-contained section.

1. Environment

The main package to install is OpenAI to access GPT-4o through an API:

```
!pip install openai==1.40.3
```

💡 **Quick tip:** Enhance your coding experience with the **AI Code Explainer** and **Quick Copy** features. Open this book in the next-gen Packt Reader. Click the **Copy** button **(1)** to quickly copy code into your coding environment, or click the **Explain** button **(2)** to get the AI assistant to explain a block of code to you.

```
                                                        Copy      Explain
function calculate(a, b) {                               1          2
  return {sum: a + b};
};
```

🔖 **The next-gen Packt Reader** is included for free with the purchase of this book. Unlock it by scanning the QR code below or visiting `https://www.packtpub.com/unlock/9781836200918`.

Make sure to freeze the OpenAI version you install. In RAG framework ecosystems, we will have to install several packages to run advanced RAG configurations. Once we have stabilized an installation, we will freeze the version of the packages installed to minimize potential conflicts between the libraries and modules we implement.

Once you have installed openai, you will have to create an account on OpenAI (if you don't have one) and obtain an API key. Make sure to check the costs and payment plans before running the API.

Once you have a key, store it in a safe place and retrieve it as follows from Google Drive, for example, as shown in the following code:

```
#API Key
#Store you key in a file and read it(you can type it directly in the notebook
but it will be visible for somebody next to you)
from google.colab import drive
drive.mount('/content/drive')
```

You can use Google Drive or any other method you choose to store your key. You can read the key from a file, or you can also choose to enter the key directly in the code:

```
f = open("drive/MyDrive/files/api_key.txt", "r")
API_KEY=f.readline().strip()
f.close()


#The OpenAI Key
import os
import openai
os.environ['OPENAI_API_KEY'] =API_KEY
openai.api_key = os.getenv("OPENAI_API_KEY")
```

With that, we have set up the main resources for our project. We will now write a generation function for the OpenAI model.

2. The generator

The code imports openai to generate content and time to measure the time the requests take:

```
import openai
from openai import OpenAI
import time
client = OpenAI()
gptmodel="gpt-4o"
start_time = time.time()   # Start timing before the request
```
We now create a function that creates a prompt with an instruction and the user input:
```
def call_llm_with_full_text(itext):
    # Join all lines to form a single string
```

```
    text_input = '\n'.join(itext)
    prompt = f"Please elaborate on the following content:\n{text_input}"
```

The function will try to call gpt-4o, adding additional information for the model:

```
    try:
      response = client.chat.completions.create(
          model=gptmodel,
          messages=[
              {"role": "system", "content": "You are an expert Natural Language
Processing exercise expert."},
              {"role": "assistant", "content": "1.You can explain read the input
and answer in detail"},
              {"role": "user", "content": prompt}
          ],
          temperature=0.1  # Add the temperature parameter here and other
parameters you need
          )
      return response.choices[0].message.content.strip()
    except Exception as e:
        return str(e)
```

Note that the instruction messages remain general in this scenario so that the model remains flexible. The temperature is low (more precise) and set to 0.1. If you wish for the system to be more creative, you can set temperature to a higher value, such as 0.7. However, in this case, it is recommended to ask for precise responses.

We can add textwrap to format the response as a nice paragraph when we call the generative AI model:

```
import textwrap
def print_formatted_response(response):
    # Define the width for wrapping the text
    wrapper = textwrap.TextWrapper(width=80)  # Set to 80 columns wide, but
adjust as needed
    wrapped_text = wrapper.fill(text=response)
    # Print the formatted response with a header and footer
    print("Response:")
    print("---------------")
    print(wrapped_text)
    print("---------------\n")
```

 The generator is now ready to be called when we need it. Due to the probabilistic nature of generative AI models, it might produce different outputs each time we call it.

The program now implements the data retrieval functionality.

3. The Data

Data collection includes text, images, audio, and video. In this notebook, we will focus on **data retrieval** through naïve, advanced, and modular configurations, not data collection. We will collect and embed data later in *Chapter 2*, *RAG Embedding Vector Stores with Deep Lake and OpenAI*. As such, we will assume that the data we need has been processed and thus collected, cleaned, and split into sentences. We will also assume that the process included loading the sentences into a Python list named db_records.

This approach illustrates three aspects of the RAG ecosystem we described in *The RAG ecosystem* section and the components of the system described in *Figure 1.3*:

- The **retriever (D)** has three **data processing** components, **collect (D1)**, **process (D2)**, and **storage (D3)**, which are preparatory phases of the retriever.
- The **retriever query (D4)** is thus independent of the first three phases (collect, process, and storage) of the retriever.
- The data processing phase will often be done independently and prior to activating the retriever query, as we will implement starting in *Chapter 2*.

This program assumes that data processing has been completed and the dataset is ready:

```
db_records = [
    "Retrieval Augmented Generation (RAG) represents a sophisticated hybrid
approach in the field of artificial intelligence, particularly within the realm
of natural language processing (NLP).",
.../...
```

We can display a formatted version of the dataset:

```
import textwrap
paragraph = ' '.join(db_records)
wrapped_text = textwrap.fill(paragraph, width=80)
print(wrapped_text)
```

The output joins the sentences in db_records for display, as printed in this excerpt, but db_records remains unchanged:

```
Retrieval Augmented Generation (RAG) represents a sophisticated hybrid approach
in the field of artificial intelligence, particularly within the realm of
natural language processing (NLP)...
```

The program is now ready to process a query.

4.The query

The **retriever (D4** in *Figure 1.3*) query process depends on how the data was processed, but the query itself is simply user input or automated input from another AI agent. We all dream of users who introduce the best input into software systems, but unfortunately, in real life, unexpected inputs lead to unpredictable behaviors. We must, therefore, build systems that take imprecise inputs into account.

In this section, we will imagine a situation in which hundreds of users in an organization have heard the word "RAG" associated with "LLM" and "vector stores." Many of them would like to understand what these terms mean to keep up with a software team that's deploying a conversational agent in their department. After a couple of days, the terms they heard become fuzzy in their memory, so they ask the conversational agent, GPT-4o in this case, to explain what they remember with the following query:

```
query = "define a rag store"
```

In this case, we will simply store the main query of the topic of this program in query, which represents the junction between the retriever and the generator. It will trigger a configuration of RAG (naïve, advanced, and modular). The choice of configuration will depend on the goals of each project.

The program takes the query and sends it to a GPT-4o model to be processed and then displays the formatted output:

```
# Call the function and print the result
llm_response = call_llm_with_full_text(query)
print_formatted_response(llm_response)
```

The output is revealing. Even the most powerful generative AI models cannot guess what a user, who knows nothing about AI, is trying to find out in good faith. In this case, GPT-4o will answer as shown in this excerpt of the output:

```
Response:
---------------
Certainly! The content you've provided appears to be a sequence of characters
that, when combined, form the phrase "define a rag store." Let's break it down
step by step:…
… This is an indefinite article used before words that begin with a consonant
sound.    - **rag**: This is a noun that typically refers to a pieceof old,
often torn, cloth.    - **store**: This is a noun that refers to a place where
goods are sold.  4. **Contextual Meaning**:    - **"Define a rag store"**:
This phrase is asking for an explanation or definition of what a "rag store"
is. 5. **Possible Definition**:     - A "rag store" could be a shop or retail
establishment that specializes in selling rags,…
```

The output will seem like a hallucination, but is it really? The user wrote the query with the good intentions of every beginner trying to learn a new topic. GPT-4o, in good faith, did what it could with the limited context it had with its probabilistic algorithm, which might even produce a different response each time we run it. However, GPT-4o is being wary of the query. It wasn't very clear, so it ends the response with the following output that asks the user for more context:

```
…Would you like more information or a different type of elaboration on this
content?…
```

The user is puzzled, not knowing what to do, and GPT-4o is awaiting further instructions. The software team has to do something!

Generative AI is based on probabilistic algorithms. As such, the response provided might vary from one run to another, providing similar (but not identical) responses.

That is when RAG comes in to save the situation. We will leave this query as it is for the whole notebook and see if a RAG-driven GPT-4o system can do better.

Part 2: Advanced techniques and evaluation

In *Part 2*, we will introduce naïve, advanced, and modular RAG. The goal is to introduce the three methods, not to process complex documents, which we will implement throughout the following chapters of this book.

Let's first begin by defining retrieval metrics to measure the accuracy of the documents we retrieve.

1. Retrieval metrics

This section explores retrieval metrics, first focusing on the role of cosine similarity in assessing the relevance of text documents. Then we will implement enhanced similarity metrics by incorporating synonym expansion and text preprocessing to improve the accuracy of similarity calculations between texts.

We will explore more metrics in the *Metrics calculation and display* section in *Chapter 7, Building Scalable Knowledge-Graph-Based RAG with Wikipedia API and LlamaIndex*.

In this chapter, let's begin with cosine similarity.

Cosine similarity

Cosine similarity measures the cosine of the angle between two vectors. In our case, the two vectors are the user query and each document in a corpus.

The program first imports the class and function we need:

```
from sklearn.feature_extraction.text import TfidfVectorizer
from sklearn.metrics.pairwise import cosine_similarity
```

TfidfVectorizer imports the class that converts text documents into a matrix of TF-IDF features. **Term Frequency-Inverse Document Frequency (TF-IDF)** quantifies the relevance of a word to a document in a collection, distinguishing common words from those significant to specific texts. TF-IDF will thus quantify word relevance in documents using frequency across the document and inverse frequency across the corpus. cosine_similarity imports the function we will use to calculate the similarity between vectors.

calculate_cosine_similarity(text1, text2) then calculates the cosine similarity between the query (text1) and each record of the dataset.

The function converts the query text (text1) and each record (text2) in the dataset into a vector with a vectorizer. Then, it calculates and returns the cosine similarity between the two vectors:

```python
def calculate_cosine_similarity(text1, text2):
    vectorizer = TfidfVectorizer(
        stop_words='english',
        use_idf=True,
        norm='l2',
        ngram_range=(1, 2),    # Use unigrams and bigrams
        sublinear_tf=True,     # Apply sublinear TF scaling
        analyzer='word'        # You could also experiment with 'char' or 'char_
wb' for character-level features
    )
    tfidf = vectorizer.fit_transform([text1, text2])
    similarity = cosine_similarity(tfidf[0:1], tfidf[1:2])
    return similarity[0][0]
```

The key parameters of this function are:

- stop_words='english: Ignores common English words to focus on meaningful content
- use_idf=True: Enables inverse document frequency weighting
- norm='l2': Applies L2 normalization to each output vector
- ngram_range=(1, 2): Considers both single words and two-word combinations
- sublinear_tf=True: Applies logarithmic term frequency scaling
- analyzer='word': Analyzes text at the word level

Cosine similarity can be limited in some cases. Cosine similarity has limitations when dealing with ambiguous queries because it strictly measures the similarity based on the angle between vector representations of text. If a user asks a vague question like "What is rag?" in the program of this chapter and the database primarily contains information on "RAG" as in "retrieval-augmented generation" for AI, not "rag cloths," the cosine similarity score might be low. This low score occurs because the mathematical model lacks contextual understanding to differentiate between the different meanings of "rag." It only computes similarity based on the presence and frequency of similar words in the text, without grasping the user's intent or the broader context of the query. Thus, even if the answers provided are technically accurate within the available dataset, the cosine similarity may not reflect the relevance accurately if the query's context isn't well-represented in the data.

In this case, we can try enhanced similarity.

Enhanced similarity

Enhanced similarity introduces calculations that leverage natural language processing tools to better capture semantic relationships between words. Using libraries like spaCy and NLTK, it preprocesses texts to reduce noise, expands terms with synonyms from WordNet, and computes similarity based on the semantic richness of the expanded vocabulary. This method aims to improve the accuracy of similarity assessments between two texts by considering a broader context than typical direct comparison methods.

The code contains four main functions:

- `get_synonyms(word)`: Retrieves synonyms for a given word from WordNet
- `preprocess_text(text)`: Converts all text to lowercase, lemmatizes gets the (roots of words), and filters stopwords (common words) and punctuation from text
- `expand_with_synonyms(words)`: Enhances a list of words by adding their synonyms
- `calculate_enhanced_similarity(text1, text2)`: Computes cosine similarity between pre-processed and synonym-expanded text vectors

The `calculate_enhanced_similarity(text1, text2)` function takes two texts and ultimately returns the cosine similarity score between two processed and synonym-expanded texts. This score quantifies the textual similarity based on their semantic content and enhanced word sets.

The code begins by downloading and importing the necessary libraries and then runs the four functions beginning with `calculate_enhanced_similarity(text1, text2)`:

```
import spacy
import nltk
nltk.download('wordnet')
from nltk.corpus import wordnet
from collections import Counter
import numpy as np


# Load spaCy model
nlp = spacy.load("en_core_web_sm")
...
```

Enhanced similarity takes this a bit further in terms of metrics. However, integrating RAG with generative AI presents multiple challenges.

No matter which metric we implement, we will face the following limitations:

- **Input versus Document Length:** User queries are often short, while retrieved documents are longer and richer, complicating direct similarity evaluations.
- **Creative Retrieval:** Systems may creatively select longer documents that meet user expectations but yield poor metric scores due to unexpected content alignment.

- **Need for Human Feedback:** Often, human judgment is crucial to accurately assess the relevance and effectiveness of retrieved content, as automated metrics may not fully capture user satisfaction. We will explore this critical aspect of RAG in *Chapter 5, Boosting RAG Performance with Expert Human Feedback*.

We will always have to find the right balance between mathematical metrics and human feedback.

We are now ready to create an example with naïve RAG.

2. Naïve RAG

Naïve RAG with keyword search and matching can prove efficient with well-defined documents within an organization, such as legal and medical documents. These documents generally have clear titles or labels for images, for example. In this naïve RAG function, we will implement keyword search and matching. To achieve this, we will apply a straightforward retrieval method in the code:

1. Split the query into individual keywords
2. Split each record in the dataset into keywords
3. Determine the length of the common matches
4. Choose the record with the best score

The generation method will:

- Augment the user input with the result of the retrieval query
- Request the generation model, which is gpt-4o in this case
- Display the response

Let's write the keyword search and matching function.

Keyword search and matching

The best matching function first initializes the best scores:

```
def find_best_match_keyword_search(query, db_records):
    best_score = 0
    best_record = None
```

The query is then split into keywords. Each record is also split into words to find the common words, measure the length of common content, and find the best match:

```
# Split the query into individual keywords
    query_keywords = set(query.lower().split())

    # Iterate through each record in db_records
    for record in db_records:
        # Split the record into keywords
        record_keywords = set(record.lower().split())
        # Calculate the number of common keywords
```

```
            common_keywords = query_keywords.intersection(record_keywords)
            current_score = len(common_keywords)
            # Update the best score and record if the current score is higher
            if current_score > best_score:
                best_score = current_score
                best_record = record


    return best_score, best_record
```

We now call the function, format the response, and print it:

```
# Assuming 'query' and 'db_records' are defined in previous cells in your Colab
notebook
best_keyword_score, best_matching_record = find_best_match_keyword_
search(query, db_records)
print(f"Best Keyword Score: {best_keyword_score}")
#print(f"Best Matching Record: {best_matching_record}")
print_formatted_response(best_matching_record)
```

The main query of this notebook will be query = "define a rag store" to see if each RAG method produces an acceptable output.

The keyword search finds the best record in the list of sentences in the dataset:

```
Best Keyword Score: 3
Response:
---------------
A RAG vector store is a database or dataset that contains vectorized data
points.
---------------
```

Let's run the metrics.

Metrics

We created the similarity metrics in the *1. Retrieval metrics* section of this chapter. We will first apply cosine similarity:

```
# Cosine Similarity
score = calculate_cosine_similarity(query, best_matching_record)
print(f"Best Cosine Similarity Score: {score:.3f}")
```

The output similarity is low, as explained in the *1. Retrieval metrics* section of this chapter. The user input is short and the response is longer and complete:

```
Best Cosine Similarity Score: 0.126
```

Enhanced similarity will produce a better score:

```
# Enhanced Similarity
response = best_matching_record
print(query,": ", response)
similarity_score = calculate_enhanced_similarity(query, response)
print(f"Enhanced Similarity:, {similarity_score:.3f}")
```

The score produced is higher with enhanced functionality:

```
define a rag store :  A RAG vector store is a database or dataset that contains
vectorized data points.
Enhanced Similarity:, 0.642
```

The output of the query will now augment the user input.

Augmented input

The augmented input is the concatenation of the user input and the best matching record of the dataset detected with the keyword search:

```
augmented_input=query+ ": "+ best_matching_record
```

The augmented input is displayed if necessary for maintenance reasons:

```
print_formatted_response(augmented_input)
```

The output then shows that the augmented input is ready:

```
Response:
---------------
define a rag store: A RAG vector store is a database or dataset that contains
vectorized data points.
--------------
```

The input is now ready for the generation process.

Generation

We are now ready to call GPT-4o and display the formatted response:

```
llm_response = call_llm_with_full_text(augmented_input)
print_formatted_response(llm_response)
```

The following excerpt of the response shows that GPT-4o understands the input and provides an interesting, pertinent response:

```
Response:
---------------
Certainly! Let's break down and elaborate on the provided content:  ### Define
a
RAG Store:  A **RAG (Retrieval-Augmented Generation) vector store** is a
specialized type of database or dataset that is designed to store and manage
vectorized data points…
```

Naïve RAG can be sufficient in many situations. However, if the volume of documents becomes too large or the content becomes more complex, then advanced RAG configurations will provide better results. Let's now explore advanced RAG.

3. Advanced RAG

As datasets grow larger, keyword search methods might prove too long to run. For instance, if we have hundreds of documents and each document contains hundreds of sentences, it will become challenging to use keyword search only. Using an index will reduce the computational load to just a fraction of the total data.

In this section, we will go beyond searching text with keywords. We will see how RAG transforms text data into numerical representations, enhancing search efficiency and processing speed. Unlike traditional methods that directly parse text, RAG first converts documents and user queries into vectors, numerical forms that speed up calculations. In simple terms, a vector is a list of numbers representing various features of text. Simple vectors might count word occurrences (term frequency), while more complex vectors, known as embeddings, capture deeper linguistic patterns.

In this section, we will implement vector search and index-based search:

- **Vector Search:** We will convert each sentence in our dataset into a numerical vector. By calculating the cosine similarity between the query vector (the user query) and these document vectors, we can quickly find the most relevant documents.
- **Index-Based Search:** In this case, all sentences are converted into vectors using **TF-IDF** (**Term Frequency-Inverse Document Frequency**), a statistical measure used to evaluate how important a word is to a document in a collection. These vectors act as indices in a matrix, allowing quick similarity comparisons without parsing each document fully.

Let's start with vector search and see these concepts in action.

3.1. Vector search

Vector search converts the user query and the documents into numerical values as vectors, enabling mathematical calculations that *retrieve relevant data faster when dealing with large volumes of data.*

The program runs through each record of the dataset to find the best matching document by computing the cosine similarity of the query vector and each record in the dataset:

```python
def find_best_match(text_input, records):
    best_score = 0
    best_record = None
    for record in records:
        current_score = calculate_cosine_similarity(text_input, record)
        if current_score > best_score:
            best_score = current_score
            best_record = record
    return best_score, best_record
```

The code then calls the vector search function and displays the best record found:

```python
best_similarity_score, best_matching_record = find_best_match(query, db_
records)
print_formatted_response(best_matching_record)
```

The output is satisfactory:

```
Response:
---------------
A RAG vector store is a database or dataset that contains vectorized data
points.
```

The response is the best one found, like with naïve RAG. This shows that there is no silver bullet. Each RAG technique has its merits. The metrics will confirm this observation.

Metrics

The metrics are the same for both similarity methods as for naïve RAG because the same document was retrieved:

```python
print(f"Best Cosine Similarity Score: {best_similarity_score:.3f}")
```

The output is:

```
Best Cosine Similarity Score: 0.126
```

And with enhanced similarity, we obtain the same output as for naïve RAG:

```python
# Enhanced Similarity
response = best_matching_record
print(query,": ", response)
similarity_score = calculate_enhanced_similarity(query, best_matching_record)
print(f"Enhanced Similarity:, {similarity_score:.3f}")
```

The output confirms the trend:

```
define a rag store :  A RAG vector store is a database or dataset that contains
vectorized data points.
Enhanced Similarity:, 0.642
```

So why use vector search if it produces the same outputs as naïve RAG? Well, in a small dataset, everything looks easy. But when we're dealing with datasets of millions of complex documents, keyword search will not capture subtleties that vectors can. Let's now augment the user query with this information retrieved.

Augmented input

We add the information retrieved to the user query with no other aid and display the result:

```
# Call the function and print the result
augmented_input=query+": "+best_matching_record
print_formatted_response(augmented_input)
```

We only added a space between the user query and the retrieved information; nothing else. The output is satisfactory:

```
Response:
---------------
define a rag store: A RAG vector store is a database or dataset that contains
vectorized data points.
---------------
```

Let's now see how the generative AI model reacts to this augmented input.

Generation

We now call GPT-4o with the augmented input and display the formatted output:

```
# Call the function and print the result
augmented_input=query+best_matching_record
llm_response = call_llm_with_full_text(augmented_input)
print_formatted_response(llm_response)
```

The response makes sense, as shown in the following excerpt:

```
Response:
---------------
Certainly! Let's break down and elaborate on the provided content:  ### Define
a RAG Store:  A **RAG (Retrieval-Augmented Generation) vector store** is a
specialized type of database or dataset that is designed to store and manage
vectorized data points…
```

While vector search significantly speeds up the process of finding relevant documents by sequentially going through each record, its efficiency can decrease as the dataset size increases. To address this scalability issue, indexed search offers a more advanced solution. Let's now see how index-based search can accelerate document retrieval.

3.2. Index-based search

Index-based search compares the vector of a user query not with the direct vector of a document's content but with an indexed vector that represents this content.

The program first imports the class and function we need:

```
from sklearn.feature_extraction.text import TfidfVectorizer
from sklearn.metrics.pairwise import cosine_similarity
```

TfidfVectorizer imports the class that converts text documents into a matrix of TF-IDF features. TF-IDF will quantify word relevance in documents using frequency across the document. The function finds the best matches using the cosine similarity function to calculate the similarity between the query and the weighted vectors of the matrix:

```
def find_best_match(query, vectorizer, tfidf_matrix):
    query_tfidf = vectorizer.transform([query])
    similarities = cosine_similarity(query_tfidf, tfidf_matrix)
    best_index = similarities.argmax()  # Get the index of the highest
similarity score
    best_score = similarities[0, best_index]
    return best_score, best_index
```

The function's main tasks are:

- **Transform Query**: Converts the input query into TF-IDF vector format using the provided vectorizer
- **Calculate Similarities**: Computes the cosine similarity between the query vector and all vectors in the tfidf_matrix
- **Identify Best Match**: Finds the index (best_index) of the highest similarity score in the results
- **Retrieve Best Score**: Extracts the highest cosine similarity score (best_score)

The output is the best similarity score found and the best index.

The following code first calls the dataset vectorizer and then searches for the most similar record through its index:

```
vectorizer, tfidf_matrix = setup_vectorizer(db_records)
best_similarity_score, best_index = find_best_match(query, vectorizer, tfidf_
matrix)
best_matching_record = db_records[best_index]
```

Finally, the results are displayed:

```
print_formatted_response(best_matching_record)
```

The system finds the best similar document to the user's input query:

```
Response:
---------------
A RAG vector store is a database or dataset that contains vectorized data
points.
---------------
```

We can see that the fuzzy user query produced a reliable output at the retrieval level before running GPT-4o.

The metrics that follow in the program are the same as for naïve and advanced RAG with vector search. This is normal because the document found is the closest to the user's input query. We will be introducing more complex documents for RAG starting in *Chapter 2, RAG Embedding Vector Stores with Deep Lake and OpenAI*. For now, let's have a look at the features that influence how the words are represented in vectors.

Feature extraction

Before augmenting the input with this document, run the following cell, which calls the `setup_vectorizer(records)` function again but displays the matrix so that you can see its format. This is shown in the following excerpt for the words "accurate" and "additional" in one of the sentences:

```
  accurate      adapt  additional
0.000000   0.000000    0.000000
0.214779   0.000000    0.000000
0.000000   0.000000    0.000000
0.000000   0.000000    0.000000
0.000000   0.000000    0.236328
```

Figure 1.4: Format of the matrix

Let's now augment the input.

Augmented input

We will simply add the query to the best matching record in a minimal way to see how GPT-4o will react and display the output:

```
augmented_input=query+": "+best_matching_record
print_formatted_response(augmented_input)
```

The output is close to or the same as with vector search, but the retrieval method is faster:

```
Response:
---------------
define a rag store: A RAG vector store is a database or dataset that contains
vectorized data points.
---------------
```

We will now plug this augmented input into the generative AI model.

Generation

We now call GPT-4o with the augmented input and display the output:

```
# Call the function and print the result
llm_response = call_llm_with_full_text(augmented_input)
print_formatted_response(llm_response)
```

The output makes sense for the user who entered the initial fuzzy query:

```
Response:
---------------
Certainly! Let's break down and elaborate on the given content:  ---  **Define
a RAG store:**  A **RAG vector store** is a **database** or **dataset** that
contains **vectorized data points**.  ---  ### Detailed Explanation:  1. **RAG
Store**:    - **RAG** stands for **Retrieval-Augmented Generation**. It is a
technique used in natural language processing (NLP) where a model retrieves
relevant information from a database or dataset to augment its generation
capabilities...
```

This approach worked well in a closed environment within an organization in a specific domain. In an open environment, the user might have to elaborate before submitting a request.

In this section, we saw that a TF-IDF matrix pre-computes document vectors, enabling faster, simultaneous comparisons without repeated vector transformations. We have seen how vector and index-based search can improve retrieval. However, in one project, we may need to apply naïve and advanced RAG depending on the documents we need to retrieve. Let's now see how modular RAG can improve our system.

4. Modular RAG

Should we use keyword search, vector search, or index-based search when implementing RAG? Each approach has its merits. The choice will depend on several factors:

- **Keyword search** suits simple retrieval
- **Vector search** is ideal for semantic-rich documents
- **Index-based search** offers speed with large data.

However, all three methods can perfectly fit together in a project. In one scenario, for example, a keyword search can help find clearly defined document labels, such as the titles of PDF files and labeled images, before they are processed. Then, indexed search will group the documents into indexed subsets. Finally, the retrieval program can search the indexed dataset, find a subset, and only use vector search to go through a limited number of documents to find the most relevant one.

In this section, we will create a `RetrievalComponent` class that can be called at each step of a project to perform the task required. The code sums up the three methods we have built in this chapter and that we can sum for the `RetrievalComponent` through its main members.

The following code initializes the class with search method choice and prepares a vectorizer if needed. `self` refers to the current instance of the class to access its variables, methods, and functions:

```python
def __init__(self, method='vector'):
        self.method = method
        if self.method == 'vector' or self.method == 'indexed':
            self.vectorizer = TfidfVectorizer()
            self.tfidf_matrix = None
```

In this case, the vector search is activated.

The `fit` method builds a TF-IDF matrix from records, and is applicable for vector or indexed search methods:

```python
def fit(self, records):
        if self.method == 'vector' or self.method == 'indexed':
            self.tfidf_matrix = self.vectorizer.fit_transform(records)
```

The retrieve method directs the query to the appropriate search method:

```python
def retrieve(self, query):
        if self.method == 'keyword':
            return self.keyword_search(query)
        elif self.method == 'vector':
            return self.vector_search(query)
        elif self.method == 'indexed':
            return self.indexed_search(query)
```

The keyword search method finds the best match by counting common keywords between queries and documents:

```python
def keyword_search(self, query):
        best_score = 0
        best_record = None
        query_keywords = set(query.lower().split())
        for index, doc in enumerate(self.documents):
            doc_keywords = set(doc.lower().split())
```

```
            common_keywords = query_keywords.intersection(doc_keywords)
            score = len(common_keywords)
            if score > best_score:
                best_score = score
                best_record = self.documents[index]
        return best_record
```

The vector search method computes similarities between query TF-IDF and document matrix and returns the best match:

```
    def vector_search(self, query):
        query_tfidf = self.vectorizer.transform([query])
        similarities = cosine_similarity(query_tfidf, self.tfidf_matrix)
        best_index = similarities.argmax()
        return db_records[best_index]
```

The indexed search method uses a precomputed TF-IDF matrix for fast retrieval of the best-matching document:

```
    def indexed_search(self, query):
        # Assuming the tfidf_matrix is precomputed and stored
        query_tfidf = self.vectorizer.transform([query])
        similarities = cosine_similarity(query_tfidf, self.tfidf_matrix)
        best_index = similarities.argmax()
        return db_records[best_index]
```

We can now activate modular RAG strategies.

Modular RAG strategies

We can call the retrieval component for any RAG configuration we wish when needed:

```
# Usage example
retrieval = RetrievalComponent(method='vector')  # Choose from 'keyword',
'vector', 'indexed'
retrieval.fit(db_records)
best_matching_record = retrieval.retrieve(query)
print_formatted_response(best_matching_record)
```

In this case, the vector search method was activated.

The following cells select the best record, as in the *3.1. Vector search* section, augment the input, call the generative model, and display the output as shown in the following excerpt:

```
Response:
----------------
Certainly! Let's break down and elaborate on the content provided:  ---
```

```
**Define a RAG store:**  A **RAG (Retrieval-Augmented Generation) store** is a
specialized type of data storage system designed to support the retrieval and
generation of information...
```

We have built a program that demonstrated how different search methodologies—keyword, vector, and index-based—can be effectively integrated into a RAG system. Each method has its unique strengths and addresses specific needs within a data retrieval context. The choice of method depends on the dataset size, query type, and performance requirements, which we will explore in the following chapters.

It's now time to summarize our explorations in this chapter and move to the next level!

Summary

RAG for generative AI relies on two main components: a retriever and a generator. The retriever processes data and defines a search method, such as fetching labeled documents with keywords—the generator's input, an LLM, benefits from augmented information when producing sequences. We went through the three main configurations of the RAG framework: naïve RAG, which accesses datasets through keywords and other entry-level search methods; advanced RAG, which introduces embeddings and indexes to improve the search methods; and modular RAG, which can combine naïve and advanced RAG as well as other ML methods.

The RAG framework relies on datasets that can contain dynamic data. A generative AI model relies on parametric data through its weights. These two approaches are not mutually exclusive. If the RAG datasets become too cumbersome, fine-tuning can prove useful. When fine-tuned models cannot respond to everyday information, RAG can come in handy. RAG frameworks also rely heavily on the ecosystem that provides the critical functionality to make the systems work. We went through the main components of the RAG ecosystem, from the retriever to the generator, for which the trainer is necessary, and the evaluator. Finally, we built an entry-level naïve, advanced, and modular RAG program in Python, leveraging keyword matching, vector search, and index-based retrieval, augmenting the input of GPT-4o.

Our next step in *Chapter 2, RAG Embedding Vector Stores with Deep Lake and OpenAI*, is to embed data in vectors. We will store the vectors in vector stores to enhance the speed and precision of the retrieval functions of a RAG ecosystem.

Questions

Answer the following questions with *Yes* or *No*:

1. Is RAG designed to improve the accuracy of generative AI models?
2. Does a naïve RAG configuration rely on complex data embedding?
3. Is fine-tuning always a better option than using RAG?
4. Does RAG retrieve data from external sources in real time to enhance responses?
5. Can RAG be applied only to text-based data?
6. Is the retrieval process in RAG triggered by a user or automated input?
7. Are cosine similarity and TF-IDF both metrics used in advanced RAG configurations?

8. Does the RAG ecosystem include only data collection and generation components?

9. Can advanced RAG configurations process multimodal data such as images and audio?

10. Is human feedback irrelevant in evaluating RAG systems?

References

- *Retrieval-Augmented Generation for Knowledge-Intensive NLP Tasks* by Patrick Lewis, Ethan Perez, Aleksandra Piktus, et al.: `https://arxiv.org/abs/2005.11401`

- *Retrieval-Augmented Generation for Large Language Models: A Survey* by Yunfan Gao, Yun Xiong, Xinyu Gao, et al.: `https://arxiv.org/abs/2312.10997`

- OpenAI models: `https://platform.openai.com/docs/models`

Further reading

- To understand why RAG-driven Generative AI transparency is recommended, please see `https://hai.stanford.edu/news/introducing-foundation-model-transparency-index`

Learn further in a live workshop with the author

If you are curious how RAG-based systems scale into agentic workflows using context engineering, consider registering for the author's live workshop on 24 January 2026, which extends the ideas discussed here into hands-on practice.

`https://packt.link/gk5Yu`

2

RAG Embedding Vector Stores with Deep Lake and OpenAI

There will come a point in the execution of your project where complexity is unavoidable when implementing RAG-driven generative AI. Embeddings transform bulky structured or unstructured texts into compact, high-dimensional vectors that capture their semantic essence, enabling faster and more efficient information retrieval. However, we will inevitably be faced with a storage issue as the creation and storage of document embeddings become necessary when managing increasingly large datasets. You could ask the question at this point, why not use keywords instead of embeddings? And the answer is simple: although embeddings require more storage space, they capture the deeper semantic meanings of texts, with more nuanced and context-aware retrieval compared to the rigid and often-matched keywords. This results in better, more pertinent retrievals. Hence, our option is to turn to vector stores in which embeddings are organized and rapidly accessible.

We will begin this chapter by exploring how to go from raw data to an Activeloop Deep Lake vector store via loading OpenAI embedding models. This requires installing and implementing several cross-platform packages, which leads us to the architecture of such systems. We will organize our RAG pipeline into separate components because breaking down the RAG pipeline into independent parts will enable several teams to work on a project simultaneously. We will then set the blueprint for a RAG-driven generative AI pipeline. Finally, we will build a three-component RAG pipeline from scratch in Python with Activeloop Deep Lake, OpenAI, and custom-built functions.

This coding journey will take us into the depths of cross-platform environment issues with packages and dependencies. We will also face the challenges of chunking data, embedding vectors, and loading them on vector stores. We will augment the input of a GPT-4o model with retrieval queries and produce solid outputs. By the end of this chapter, you will fully understand how to leverage the power of embedded documents in vector stores for generative AI.

To sum up, this chapter covers the following topics:

- Introducing document embeddings and vector stores
- How to break a RAG pipeline into independent components

- Building a RAG pipeline from raw data to Activeloop Deep Lake
- Facing the environmental challenge of cross-platform packages and libraries
- Leveraging the power of LLMs to embed data with an OpenAI embedding model
- Querying an Activeloop Deep Lake vector store to augment user inputs
- Generative solid augmented outputs with OpenAI GPT-4o

Let's begin by learning how to go from raw data to a vector store.

From raw data to embeddings in vector stores

Embeddings convert any form of data (text, images, or audio) into real numbers. Thus, a document is converted into a vector. These mathematical representations of documents allow us to calculate the distances between documents and retrieve similar data.

The raw data (books, articles, blogs, pictures, or songs) is first collected and cleaned to remove noise. The prepared data is then fed into a model such as OpenAI `text-embedding-3-small`, which will embed the data. Activeloop Deep Lake, for example, which we will implement in this chapter, will break a text down into pre-defined chunks defined by a certain number of characters. The size of a chunk could be 1,000 characters, for instance. We can let the system optimize these chunks, as we will implement them in the *Optimizing chunking* section of the next chapter. These chunks of text make it easier to process large amounts of data and provide more detailed embeddings of a document, as shown here:

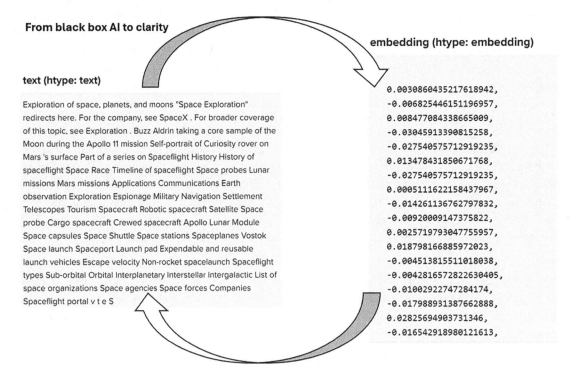

Figure 2.1: Excerpt of an Activeloop vector store dataset record

Transparency has been the holy grail in AI since the beginning of parametric models, in which the information is buried in learned parameters that produce black box systems. RAG is a game changer, as shown in *Figure 2.1*, because the content is fully traceable:

- Left side (Text): In RAG frameworks, every piece of generated content is traceable back to its source data, ensuring the output's transparency. The OpenAI generative model will respond, taking the augmented input into account.

- Right side (Embeddings): Data embeddings are directly visible and linked to the text, contrasting with parametric models where data origins are encoded within model parameters.

Once we have our text and embeddings, the next step is to store them efficiently for quick retrieval. This is where *vector stores* come into play. A vector store is a specialized database designed to handle high-dimensional data like embeddings. We can create datasets on serverless platforms such as Activeloop, as shown in *Figure 2.2*. We can create and access them in code through an API, as we will do in the *Building a RAG pipeline* section of this chapter.

Figure 2.2: Managing datasets with vector stores

Another feature of vector stores is their ability to retrieve data with optimized methods. Vector stores are built with powerful indexing methods, which we will discuss in the next chapter. This retrieving capacity allows a RAG model to quickly find and retrieve the most relevant embeddings during the generation phase, augment user inputs, and increase the model's ability to produce high-quality output.

We will now see how to organize a RAG pipeline that goes from data collection, processing, and retrieval to augmented-input generation.

Organizing RAG in a pipeline

A RAG pipeline will typically collect data and prepare it by cleaning it, for example, chunking the documents, embedding them, and storing them in a vector store dataset. The vector dataset is then queried to augment the user input of a generative AI model to produce an output. However, it is highly recommended not to run this sequence of RAG in one single program when it comes to using a vector store. We should at least separate the process into three components:

- Data collection and preparation

- Data embedding and loading into the dataset of a vector store
- Querying the vectorized dataset to augment the input of a generative AI model to produce a response

Let's go through the main reasons for this component approach:

- **Specialization**, which will allow each member of a team to do what they are best at, either collecting and cleaning data, running embedding models, managing vector stores, or tweaking generative AI models.
- **Scalability**, making it easier to upgrade separate components as the technology evolves and scale the different components with specialized methods. Storing raw data, for example, can be scaled on a different server than the cloud platform, where the embedded vectors are stored in a vectorized dataset.
- **Parallel development**, which allows each team to advance at their pace without waiting for others. Improvements can be made continually on one component without disrupting the processes of the other components.
- **Maintenance** is component-independent. One team can work on one component without affecting the other parts of the system. For example, if the RAG pipeline is in production, users can continue querying and running generative AI through the vector store while a team fixes the data collection component.
- **Security** concerns and privacy are minimized because each team can work separately with specific authorization, access, and roles for each component.

As we can see, in real-life production environments or large-scale projects, it is rare for a single program or team to manage end-to-end processes. We are now ready to draw the blueprint of the RAG pipeline that we will build in Python in this chapter.

A RAG-driven generative AI pipeline

Let's dive into what a real-life RAG pipeline looks like. Imagine we're a team that has to deliver a whole system in just a few weeks. Right off the bat, we're bombarded with questions like:

- Who's going to gather and clean up all the data?
- Who's going to handle setting up OpenAI's embedding model?
- Who's writing the code to get those embeddings up and running and managing the vector store?
- Who's going to take care of implementing GPT-4 and managing what it spits out?

Within a few minutes, everyone starts looking pretty worried. The whole thing feels overwhelming—like, seriously, who would even think about tackling all that alone?

So here's what we do. We split into three groups, each of us taking on different parts of the pipeline, as shown in *Figure 2.3*:

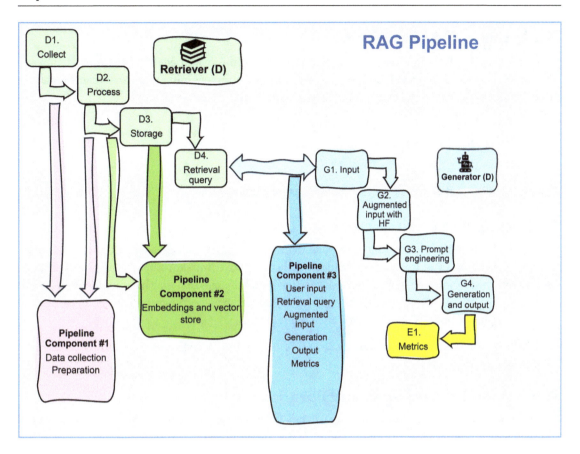

Figure 2.3: RAG pipeline components

Each of the three groups has one component to implement:

- **Data Collection and Prep (D1 and D2):** One team takes on collecting the data and cleaning it.
- **Data Embedding and Storage (D2 and D3):** Another team works on getting the data through OpenAI's embedding model and stores these vectors in an Activeloop Deep Lake dataset.
- **Augmented Generation (D4, G1-G4, and E1):** The last team handles the big job of generating content based on user input and retrieval queries. They use GPT-4 for this, and even though it sounds like a lot, it's actually a bit easier because they aren't waiting on anyone else—they just need the computer to do its calculations and evaluate the output.

Suddenly, the project doesn't seem so scary. Everyone has their part to focus on, and we can all work without being distracted by the other teams. This way, we can all move faster and get the job done without the hold-ups that usually slow things down.

The organization of the project, represented in *Figure 2.3*, is a variant of the RAG ecosystem's framework represented in *Figure 1.3* of *Chapter 1, Why Retrieval Augmented Generation?*

We can now begin building a RAG pipeline.

Building a RAG pipeline

We will now build a RAG pipeline by implementing the pipeline described in the previous section and illustrated in *Figure 2.3*. We will implement three components assuming that three teams (Team #1, Team #2, and Team #3) work in parallel to implement the pipeline:

- Data collection and preparation by Team #1
- Data embedding and storage by Team #2
- Augmented generation by Team #3

The first step is to set up the environment for these components.

Setting up the environment

Let's face it here and now. Installing cross-platform, cross-library packages with their dependencies can be quite challenging! It is important to take this complexity into account and be prepared to get the environment running correctly. Each package has dependencies that may have conflicting versions. Even if we adapt the versions, an application may not run as expected anymore. So, take your time to install the right versions of the packages and dependencies.

We will only describe the environment once in this section for all three components and refer to this section when necessary.

The installation packages and libraries

To build the RAG pipeline in this section, we will need packages and need to freeze the package versions to prevent dependency conflicts and issues with the functions of the libraries, such as:

- Possible conflicts between the versions of the dependencies.
- Possible conflicts when one of the libraries needs to be updated for an application to run. For example, in August 2024, installing Deep Lake required Pillow version 10.x.x and Google Colab's version was 9.x.x. Thus, it was necessary to uninstall Pillow and reinstall it with a recent version before installing Deep Lake. Google Colab will no doubt update Pillow. Many cases such as this occur in a fast-moving market.
- Possible deprecations if the versions remain frozen for too long.
- Possible issues if the versions are frozen for too long and bugs are not corrected by upgrades.

Thus, if we freeze the versions, an application may remain stable for some time but encounter issues. But if we upgrade the versions too quickly, some of the other libraries may not work anymore. There is no silver bullet! It's a continual quality control process.

For our program, in this section, we will freeze the versions. Let's now go through the installation steps to create the environment for our pipeline.

The components involved in the installation process

Let's begin by describing the components that are installed in the *Installing the environment* section of each notebook. The components are not necessarily installed in all notebooks; this section serves as an inventory of the packages.

In the first pipeline section, *1. Data collection and preparation*, we will only need to install Beautiful Soup and Requests:

```
!pip install beautifulsoup4==4.12.3
!pip install requests==2.31.0
```

This explains why this component of the pipeline should remain separate. It's a straightforward job for a developer who enjoys creating interfaces to interact with the web. It's also a perfect fit for a junior developer who wants to get involved in data collection and analysis.

The two other pipeline components we will build in this section, *2. Data embedding and storage* and *3. Augmented generation*, will require more attention as well as the installation of requirements01.txt, as explained in the previous section. For now, let's continue with the installation step by step.

Mounting a drive

In this scenario, the program mounts Google Drive in Google Colab to safely read the OpenAI API key to access OpenAI models and the Activeloop API token for authentication to access Activeloop Deep Lake datasets:

```
#Google Drive option to store API Keys
#Store your key in a file and read it(you can type it directly in the #
#notebook but it will be visible for somebody next to you)
from google.colab import drive
drive.mount('/content/drive')
```

You can choose to store your keys and tokens elsewhere. Just make sure they are in a safe location.

Creating a subprocess to download files from GitHub

The goal here is to write a function to download the grequests.py file from GitHub. This program contains a function to download files using curl, with the option to add a private token if necessary:

```
import subprocess
url = "https://raw.githubusercontent.com/Denis2054/RAG-Driven-Generative-AI/
main/commons/grequests.py"
output_file = "grequests.py"

# Prepare the curl command using the private token
curl_command = [
    "curl",
    "-o", output_file,
    url
]

# Execute the curl command
try:
```

```
    subprocess.run(curl_command, check=True)
    print("Download successful.")
except subprocess.CalledProcessError:
    print("Failed to download the file.")
```

The grequests.py file contains a function that can, if necessary, accept a private token or any other security system that requires credentials when retrieving data with curl commands:

```
import subprocess
import os

# add a private token after the filename if necessary
def download(directory, filename):
    # The base URL of the image files in the GitHub repository
    base_url = 'https://raw.githubusercontent.com/Denis2054/RAG-Driven-
Generative-AI/main/'

    # Complete URL for the file
    file_url = f"{base_url}{directory}/{filename}"

    # Use curl to download the file, including an Authorization header for the
private token
    try:
        # Prepare the curl command with the Authorization header
        #curl_command = f'curl -H "Authorization: token {private_token}" -o
{filename} {file_url}'
        curl_command = f'curl -H -o {filename} {file_url}'

        # Execute the curl command
        subprocess.run(curl_command, check=True, shell=True)
        print(f"Downloaded '{filename}' successfully.")
    except subprocess.CalledProcessError:
        print(f"Failed to download '{filename}'. Check the URL, your internet
connection, and if the token is correct and has appropriate permissions.")
```

Installing requirements

Now, we will install the requirements for this section when working with Activeloop Deep Lake and OpenAI. We will only need:

```
!pip install deeplake==3.9.18
!pip install openai==1.40.3
```

As of August 2024, Google Colab's version of Pillow conflicts with deeplake's package. However, the deeplake installation package deals with this automatically. All you have to do is restart the session and run it again, which is why pip install deeplake==3.9.18 is the first line of each notebook it is installed in.

After installing the requirements, we must run a line of code for Activeloop to activate a public DNS server:

```
# For Google Colab and Activeloop(Deeplake library)
#This line writes the string "nameserver 8.8.8.8" to the file. This is
specifying that the DNS server the system
#should use is at the IP address 8.8.8.8, which is one of Google's Public DNS
servers.
with open('/etc/resolv.conf', 'w') as file:
    file.write("nameserver 8.8.8.8")
```

Authentication process

You will need to sign up to OpenAI to obtain an API key: https://openai.com/. Make sure to check the pricing policy before using the key. First, let's activate OpenAI's API key:

```
#Retrieving and setting OpenAI API key
f = open("drive/MyDrive/files/api_key.txt", "r")
API_KEY=f.readline().strip()
f.close()


#The OpenAI API key
import os
import openai
os.environ['OPENAI_API_KEY'] =API_KEY
openai.api_key = os.getenv("OPENAI_API_KEY")
```

Then, we activate Activeloop's API token for Deep Lake:

```python
#Retrieving and setting Activeloop API token
f = open("drive/MyDrive/files/activeloop.txt", "r")
API_token=f.readline().strip()
f.close()
ACTIVELOOP_TOKEN=API_token
os.environ['ACTIVELOOP_TOKEN'] =ACTIVELOOP_TOKEN
```

You will need to sign up on Activeloop to obtain an API token: https://www.activeloop.ai/. Again, make sure to check the pricing policy before using the Activeloop token.

Once the environment is installed, you can hide the *Installing the environment* cells we just ran to focus on the content of the pipeline components, as shown in *Figure 2.4*:

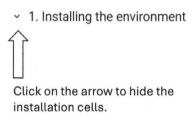

Figure 2:4: Hiding the installation cells

The installation cells will then be hidden but can still be run, as shown in *Figure 2.5*:

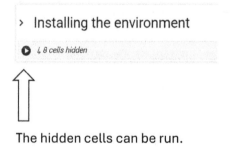

Figure 2.5: Running hidden cells

We can now focus on the pipeline components for each pipeline component. Let's begin with data collection and preparation.

1. Data collection and preparation

Data collection and preparation is the first pipeline component, as described earlier in this chapter. Team #1 will only focus on their component, as shown in *Figure 2.6*:

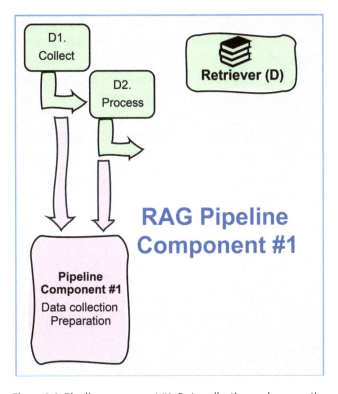

Figure 2.6: Pipeline component #1: Data collection and preparation

Let's jump in and lend a hand to `Team #1`. Our work is clearly defined, so we can enjoy the time taken to implement the component. We will retrieve and process 10 Wikipedia articles that provide a comprehensive view of various aspects of space exploration:

- **Space exploration**: Overview of the history, technologies, missions, and plans involved in the exploration of space (`https://en.wikipedia.org/wiki/Space_exploration`)

- **Apollo program**: Details about the NASA program that landed the first humans on the Moon and its significant missions (`https://en.wikipedia.org/wiki/Apollo_program`)

- **Hubble Space Telescope**: Information on one of the most significant telescopes ever built, which has been crucial in many astronomical discoveries (`https://en.wikipedia.org/wiki/Hubble_Space_Telescope`)

- **Mars rover**: Insight into the rovers that have been sent to Mars to study its surface and environment (`https://en.wikipedia.org/wiki/Mars_rover`)

- **International Space Station (ISS)**: Details about the ISS, its construction, international collaboration, and its role in space research (`https://en.wikipedia.org/wiki/International_Space_Station`)

- **SpaceX**: Covers the history, achievements, and goals of SpaceX, one of the most influential private spaceflight companies (`https://en.wikipedia.org/wiki/SpaceX`)

- **Juno (spacecraft)**: Information about the NASA space probe that orbits and studies Jupiter, its structure, and moons (`https://en.wikipedia.org/wiki/Juno_(spacecraft)`)

- **Voyager program**: Details on the Voyager missions, including their contributions to our understanding of the outer solar system and interstellar space (`https://en.wikipedia.org/wiki/Voyager_program`)
- **Galileo (spacecraft)**: Overview of the mission that studied Jupiter and its moons, providing valuable data on the gas giant and its system (`https://en.wikipedia.org/wiki/Galileo_(spacecraft)`)
- **Kepler space telescope**: Information about the space telescope designed to discover Earth-size planets orbiting other stars (`https://en.wikipedia.org/wiki/Kepler_Space_Telescope`)

These articles cover a wide range of topics in space exploration, from historical programs to modern technological advances and missions.

Now, open `1-Data_collection_preparation.ipynb` in the GitHub repository. We will first collect the data.

Collecting the data

We just need `import requests` for the HTTP requests, `from bs4 import BeautifulSoup` for HTML parsing, and `import re`, the regular expressions module:

```
import requests
from bs4 import BeautifulSoup
import re
```

We then select the URLs we need:

```
# URLs of the Wikipedia articles
urls = [
    "https://en.wikipedia.org/wiki/Space_exploration",
    "https://en.wikipedia.org/wiki/Apollo_program",
    "https://en.wikipedia.org/wiki/Hubble_Space_Telescope",
    "https://en.wikipedia.org/wiki/Mars_over",
    "https://en.wikipedia.org/wiki/International_Space_Station",
    "https://en.wikipedia.org/wiki/SpaceX",
    "https://en.wikipedia.org/wiki/Juno_(spacecraft)",
    "https://en.wikipedia.org/wiki/Voyager_program",
    "https://en.wikipedia.org/wiki/Galileo_(spacecraft)",
    "https://en.wikipedia.org/wiki/Kepler_Space_Telescope"
]
```

This list is in code. However, it could be stored in a database, a file, or any other format, such as JSON. We can now prepare the data.

Preparing the data

First, we write a cleaning function. This function removes numerical references such as [1] [2] from a given text string, using regular expressions, and returns the cleaned text:

```python
def clean_text(content):
    # Remove references that usually appear as [1], [2], etc.
    content = re.sub(r'\[\d+\]', '', content)
    return content
```

Then, we write a classical fetch and clean function, which will return a nice and clean text by extracting the content we need from the documents:

```python
def fetch_and_clean(url):
    # Fetch the content of the URL
    response = requests.get(url)
    soup = BeautifulSoup(response.content, 'html.parser')

    # Find the main content of the article, ignoring side boxes and headers
    content = soup.find('div', {'class': 'mw-parser-output'})

    # Remove the bibliography section, which generally follows a header like
"References", "Bibliography"
    for section_title in ['References', 'Bibliography', 'External links', 'See
also']:
        section = content.find('span', id=section_title)
        if section:
            # Remove all content from this section to the end of the document
            for sib in section.parent.find_next_siblings():
                sib.decompose()
            section.parent.decompose()

    # Extract and clean the text
    text = content.get_text(separator=' ', strip=True)
    text = clean_text(text)
    return text
```

Finally, we write the content in llm.txt file for the team working on the data embedding and storage functions:

```
# File to write the clean text
with open('llm.txt', 'w', encoding='utf-8') as file:
    for url in urls:
        clean_article_text = fetch_and_clean(url)
        file.write(clean_article_text + '\n')


print("Content written to llm.txt")
```

The output confirms that the text has been written:

```
Content written to llm.txt
```

The program can be modified to save the data in other formats and locations, as required for a project's specific needs. The file can then be verified before we move on to the next batch of data to retrieve and process:

```
# Open the file and read the first 20 lines
with open('llm.txt', 'r', encoding='utf-8') as file:
    lines = file.readlines()
    # Print the first 20 lines
    for line in lines[:20]:
        print(line.strip())
```

The output shows the first lines of the document that will be processed:

```
Exploration of space, planets, and moons "Space Exploration" redirects
here. For the company, see SpaceX . For broader coverage of this topic, see
Exploration . Buzz Aldrin taking a core sample of the Moon during the Apollo 11
mission…
```

This component can be managed by a team that enjoys searching for documents on the web or within a company's data environment. The team will gain experience in identifying the best documents for a project, which is the foundation of any RAG framework.

Team #2 can now work on the data to embed the documents and store them.

2. Data embedding and storage

Team #2's job is to focus on the second component of the pipeline. They will receive batches of prepared data to work on. They don't have to worry about retrieving data. Team #1 has their back with their data collection and preparation component.

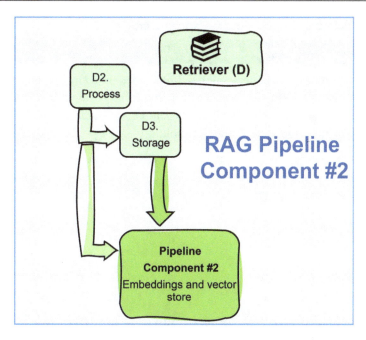

Figure 2.7: Pipeline component #2: Data embedding and storage

Let's now jump in and help Team #2 to get the job done. Open 2-Embeddings_vector_store.ipynb in the GitHub Repository. We will embed and store the data provided by Team #1 and retrieve a batch of documents to work on.

Retrieving a batch of prepared documents

First, we download a batch of documents available on a server and provided by Team #1, which is the first of a continual stream of incoming documents. In this case, we assume it's the space exploration file:

```
from grequests import download
source_text = "llm.txt"

directory = "Chapter02"
filename = "llm.txt"
download(directory, filename)
```

Note that source_text = "llm.txt" will be used by the function that will add the data to our vector store. We then briefly check the document just to be sure, knowing that Team #1 has already verified the information:

```
# Open the file and read the first 20 lines
with open('llm.txt', 'r', encoding='utf-8') as file:
    lines = file.readlines()
    # Print the first 20 lines
    for line in lines[:20]:
        print(line.strip())
```

The output is satisfactory, as shown in the following excerpt:

```
Exploration of space, planets, and moons "Space Exploration" redirects here.
```

We will now chunk the data. We will determine a chunk size defined by the number of characters. In this case, it is CHUNK_SIZE = 1000, but we can select chunk sizes using different strategies. *Chapter 7, Building Scalable Knowledge-Graph-based RAG with Wikipedia API and LlamaIndex*, will take chunk size optimization further with automated seamless chunking.

Chunking is necessary to optimize data processing: selecting portions of text, embedding, and loading the data. It also makes the embedded dataset easier to query. The following code chunks a document to complete the preparation process:

```
with open(source_text, 'r') as f:
    text = f.read()


CHUNK_SIZE = 1000
chunked_text = [text[i:i+CHUNK_SIZE] for i in range(0,len(text), CHUNK_SIZE)]
```

We are now ready to create a vector store to vectorize data or add data to an existing one.

Verifying if the vector store exists and creating it if not

First, we need to define the path of our Activeloop vector store path, whether our dataset exists or not:

```
vector_store_path = "hub://denis76/space_exploration_v1"
```

 Make sure to replace `hub://denis76/space_exploration_v1` with your organization and dataset name.

Then, we write a function to attempt to load the vector store or automatically create one if it doesn't exist:

```
from deeplake.core.vectorstore.deeplake_vectorstore import VectorStore
import deeplake.util

try:
    # Attempt to load the vector store
    vector_store = VectorStore(path=vector_store_path)
    print("Vector store exists")
except FileNotFoundError:
    print("Vector store does not exist. You can create it.")
    # Code to create the vector store goes here
    create_vector_store=True
```

The output confirms that the vector store has been created:

```
Your Deep Lake dataset has been successfully created!
Vector store exists
```

We now need to create an embedding function.

The embedding function

The embedding function will transform the chunks of data we created into vectors to enable vector-based search. In this program, we will use "text-embedding-3-small" to embed the documents.

OpenAI has other embedding models that you can use: https://platform.openai.com/docs/models/ embeddings. *Chapter 6, Scaling RAG Bank Customer Data with Pinecone*, provides alternative code for embedding models in the *Embedding* section. In any case, it is recommended to evaluate embedding models before choosing one in production. Examine the characteristics of each embedding model, as described by OpenAI, focusing on their length and capacities. text-embedding-3-small was chosen in this case because it stands out as a robust choice for efficiency and speed:

```
def embedding_function(texts, model="text-embedding-3-small"):
    if isinstance(texts, str):
        texts = [texts]
    texts = [t.replace("\n", " ") for t in texts]
    return [data.embedding for data in openai.embeddings.create(input = texts,
model=model).data]
```

The text-embedding-3-small text embedding model from OpenAI typically uses embeddings with a restricted number of dimensions, to balance obtaining enough detail in the embeddings with large computational workloads and storage space. Make sure to check the model page and pricing information before running the code: https://platform.openai.com/docs/guides/embeddings/embedding-models.

We are now all set to begin populating the vector store.

Adding data to the vector store

We set the adding data flag to True:

```
add_to_vector_store=True
if add_to_vector_store == True:
    with open(source_text, 'r') as f:
        text = f.read()
        CHUNK_SIZE = 1000
        chunked_text = [text[i:i+1000] for i in range(0, len(text), CHUNK_
SIZE)]

vector_store.add(text = chunked_text,
            embedding_function = embedding_function,
```

```
                      embedding_data = chunked_text,
                      metadata = [{"source": source_text}]*len(chunked_text))
```

The source text, `source_text = "llm.txt"`, has been embedded and stored. A summary of the dataset's structure is displayed, showing that the dataset was loaded:

```
Creating 839 embeddings in 2 batches of size 500:: 100%|████████| 2/2
[01:44<00:00, 52.04s/it]
Dataset(path='hub://denis76/space_exploration_v1', tensors=['text', 'metadata',
'embedding', 'id'])

  tensor       htype        shape      dtype  compression
  -------     -------      -------    -------  -------
   text        text       (839, 1)      str      None
  metadata     json       (839, 1)      str      None
  embedding  embedding  (839, 1536)  float32     None
    id         text       (839, 1)      str      None
```

Observe that the dataset contains four tensors:

*   `embedding`: Each chunk of data is embedded in a vector
*   `id`: The ID is a string of characters and is unique
*   `metadata`: The metadata contains the source of the data—in this case, the `llm.txt` file.
*   `text`: The content of a chunk of text in the dataset

This dataset structure can vary from one project to another, as we will see in *Chapter 4, Multimodal Modular RAG for Drone Technology*. We can also visualize how the dataset is organized at any time to verify the structure. The following code will display the summary that was just displayed:

```
# Print the summary of the Vector Store
print(vector_store.summary())
```

We can also visualize vector store information if we wish.

Vector store information

Activeloop's API reference provides us with all the information we need to manage our datasets: `https://docs.deeplake.ai/en/latest/`. We can visualize our datasets once we sign in at `https://app.activeloop.ai/datasets/mydatasets/`.

We can also load our dataset in one line of code:

```
ds = deeplake.load(vector_store_path)
```

The output provides a path to visualize our datasets and query and explore them online:

```
This dataset can be visualized in Jupyter Notebook by ds.visualize() or at
https://app.activeloop.ai/denis76/space_exploration_v1
hub://denis76/space_exploration_v1 loaded successfully.
```

You can also access your dataset directly on Activeloop by signing in and going to your datasets. You will find online dataset exploration tools to query your dataset and more, as shown here:

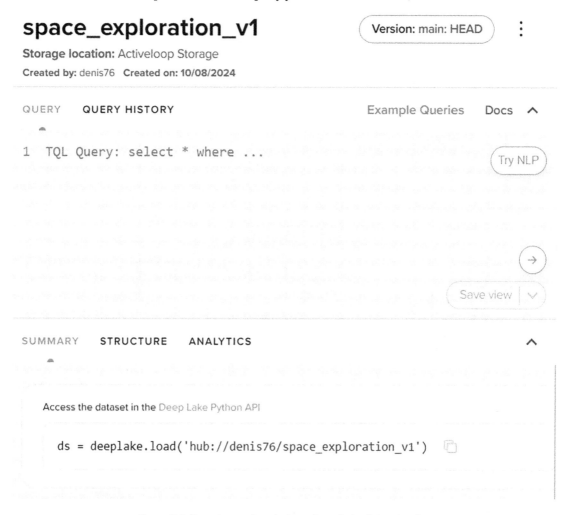

Figure 2.8: Querying and exploring a Deep Lake dataset online.

Among the many functions available, we can display the estimated size of a dataset:

```
#Estimates the size in bytes of the dataset.
ds_size=ds.size_approx()
```

Once we have obtained the size, we can convert it into megabytes and gigabytes:

```
# Convert bytes to megabytes and limit to 5 decimal places
ds_size_mb = ds_size / 1048576
print(f"Dataset size in megabytes: {ds_size_mb:.5f} MB")


# Convert bytes to gigabytes and limit to 5 decimal places
```

```
ds_size_gb = ds_size / 1073741824
print(f"Dataset size in gigabytes: {ds_size_gb:.5f} GB")
```

The output shows the size of the dataset in megabytes and gigabytes:

```
Dataset size in megabytes: 55.31311 MB
Dataset size in gigabytes: 0.05402 GB
```

Team #2's pipeline component for data embedding and storage seems to be working. Let's now explore augmented generation.

3. Augmented input generation

Augmented generation is the third pipeline component. We will use the data we retrieved to augment the user input. This component processes the user input, queries the vector store, augments the input, and calls gpt-4-turbo, as shown in *Figure 2.9*:

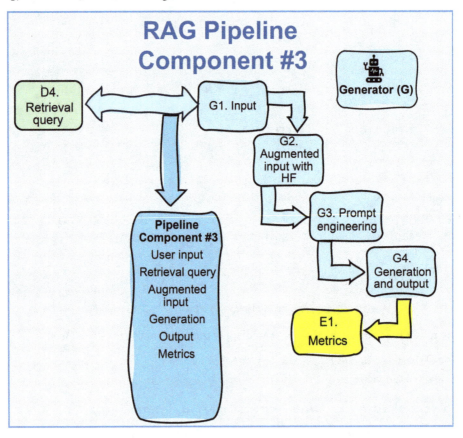

Figure 2.9: Pipeline component #3: Augmented input generation

Figure 2.9 shows that pipeline component #3 fully deserves its **Retrieval Augmented Generation (RAG)** name. However, it would be impossible to run this component without the work put in by Team #1 and Team #2 to provide the necessary information to generate augmented input content.

Let's jump in and see how Team #3 does the job. Open 3-Augmented_Generation.ipynb in the GitHub repository. The *Installing the environment* section of the notebook is described in the *Setting up the environment* section of this chapter. We select the vector store (replace the vector store path with your vector store):

```
vector_store_path = "hub://denis76/space_exploration_v1"
```

Then, we load the dataset:

```
from deeplake.core.vectorstore.deeplake_vectorstore import VectorStore
import deeplake.util
ds = deeplake.load(vector_store_path)
```

We print a confirmation message that the vector store exists. At this point stage, Team #2 previously ensured that everything was working well, so we can just move ahead rapidly:

```
vector_store = VectorStore(path=vector_store_path)
```

The output confirms that the dataset exists and is loaded:

```
Deep Lake Dataset in hub://denis76/space_exploration_v1 already exists, loading
from the storage
```

We assume that pipeline component #2, as built in the *Data embedding and storage* section, has created and populated the vector_store and has verified that it can be queried. Let's now process the user input.

Input and query retrieval

We will need the embedding function to embed the user input:

```
def embedding_function(texts, model="text-embedding-3-small"):
    if isinstance(texts, str):
        texts = [texts]
    texts = [t.replace("\n", " ") for t in texts]
    return [data.embedding for data in openai.embeddings.create(input = texts,
model=model).data]
```

Note that we are using the same embedding model as the data embedding and storage component to ensure full compatibility between the input and the vector dataset: text-embedding-ada-002.

We can now either use an interactive prompt for an input or process user inputs in batches. In this case, we process a user input that has already been entered that could be fetched from a user interface, for example.

We first ask the user for an input or define one:

```
def get_user_prompt():
    # Request user input for the search prompt
    return input("Enter your search query: ")
```

```
# Get the user's search query
#user_prompt = get_user_prompt()
user_prompt="Tell me about space exploration on the Moon and Mars."
```

We then plug the prompt into the search query and store the output in `search_results`:

```
search_results = vector_store.search(embedding_data=user_prompt, embedding_
function=embedding_function)
```

The user prompt and search results stored in `search_results` are formatted to be displayed. First, let's print the user prompt:

```
print(user_prompt)
```

We can also wrap the retrieved text to obtain a formatted output:

```
# Function to wrap text to a specified width
def wrap_text(text, width=80):
    lines = []
    while len(text) > width:
        split_index = text.rfind(' ', 0, width)
        if split_index == -1:
            split_index = width
        lines.append(text[:split_index])
        text = text[split_index:].strip()
    lines.append(text)
    return '\n'.join(lines)
```

However, let's only select one of the top results and print it:

```
import textwrap

# Assuming the search results are ordered with the top result first
top_score = search_results['score'][0]
top_text = search_results['text'][0].strip()
top_metadata = search_results['metadata'][0]['source']

# Print the top search result
print("Top Search Result:")
print(f"Score: {top_score}")
print(f"Source: {top_metadata}")
print("Text:")
print(wrap_text(top_text))
```

The following output shows that we have a reasonably good match:

```
Top Search Result:
Score: 0.6016581654548645
Source: llm.txt
Text:
Exploration of space, planets, and moons "Space Exploration" redirects here.
For the company, see SpaceX . For broader coverage of this topic, see
Exploration . Buzz Aldrin taking a core sample of the Moon during the Apollo 11
mission Self-portrait of Curiosity rover on Mars 's surface Part of a series
on…
```

We are ready to augment the input with the additional information we have retrieved.

Augmented input

The program adds the top retrieved text to the user input:

```
augmented_input=user_prompt+" "+top_text
print(augmented_input)
```

The output displays the augmented input:

```
Tell me about space exploration on the Moon and Mars. Exploration of space,
planets …
```

gpt-4o can now process the augmented input and generate content:

```
from openai import OpenAI
client = OpenAI()
import time

gpt_model = "gpt-4o"
start_time = time.time()  # Start timing before the request
```

Note that we are timing the process. We now write the generative AI call, adding roles to the message we create for the model:

```
def call_gpt4_with_full_text(itext):
    # Join all lines to form a single string
    text_input = '\n'.join(itext)
    prompt = f"Please summarize or elaborate on the following content:\n{text_
input}"

    try:
        response = client.chat.completions.create(
            model=gpt_model,
```

```
            messages=[
                {"role": "system", "content": "You are a space exploration
    expert."},
                {"role": "assistant", "content": "You can read the input and
    answer in detail."},
                {"role": "user", "content": prompt}
            ],
            temperature=0.1  # Fine-tune parameters as needed
        )
        return response.choices[0].message.content.strip()
    except Exception as e:
        return str(e)
```

The generative model is called with the augmented input; the response time is calculated and displayed along with the output:

```
gpt4_response = call_gpt4_with_full_text(augmented_input)

response_time = time.time() - start_time  # Measure response time
print(f"Response Time: {response_time:.2f} seconds")  # Print response time

print(gpt_model, "Response:", gpt4_response)
```

Note that the raw output is displayed with the response time:

```
Response Time: 8.44 seconds
gpt-4o Response: Space exploration on the Moon and Mars has been a significant
focus of human spaceflight and robotic missions. Here's a detailed summary…
```

Let's format the output with `textwrap` and print the result. `print_formatted_response(response)` first checks if the response returned contains Markdown features. If so, it will format the response; if not, it will perform a standard output text wrap:

```
import textwrap
import re
from IPython.display import display, Markdown, HTML
import markdown

def print_formatted_response(response):
    # Check for markdown by looking for patterns like headers, bold, lists,
etc.
    markdown_patterns = [
        r"^#+\s",              # Headers
        r"^\*+",               # Bullet points
        r"\*\*",               # Bold
```

```
        r"_",              # Italics
        r"\[.+\]\(.+\)",   # Links
        r"-\s",            # Dashes used for lists
        r"\`\`\`"          # Code blocks
    ]


    # If any pattern matches, assume the response is in markdown
    if any(re.search(pattern, response, re.MULTILINE) for pattern in markdown_
patterns):
        # Markdown detected, convert to HTML for nicer display
        html_output = markdown.markdown(response)
        display(HTML(html_output))  # Use display(HTML()) to render HTML in
Colab
    else:
        # No markdown detected, wrap and print as plain text
        wrapper = textwrap.TextWrapper(width=80)
        wrapped_text = wrapper.fill(text=response)

        print("Text Response:")
        print("--------------------")
        print(wrapped_text)
        print("-------------------\n")

print_formatted_response(gpt4_response)
```

The output is satisfactory:

```
Moon Exploration
    Historical Missions:
    1. Apollo Missions: NASA's Apollo program, particularly Apollo 11, marked
the first manned Moon landing in 1969. Astronauts like Buzz Aldrin collected
core samples and conducted experiments.
    2. Lunar Missions: Various missions have been conducted to explore the
Moon, including robotic landers and orbiters from different countries.
Scientific Goals:
    3. Geological Studies: Understanding the Moon's composition, structure, and
history.
    4. Resource Utilization: Investigating the potential for mining resources
like Helium-3 and water ice.
    Future Plans:
    1. Artemis Program: NASA's initiative to return humans to the Moon and
establish a sustainable presence by the late 2020s.
```

```
    2. International Collaboration: Partnerships with other space agencies and
private companies to build lunar bases and conduct scientific research.

Mars Exploration
    Robotic Missions:
    1. Rovers: NASA's rovers like Curiosity and Perseverance have been
exploring Mars' surface, analyzing soil and rock samples, and searching for
signs of past life.
    2. Orbiters: Various orbiters have been mapping Mars' surface and studying
its atmosphere…
```

Let's introduce an evaluation metric to measure the quality of the output.

Evaluating the output with cosine similarity

In this section, we will implement cosine similarity to measure the similarity between user input and the generative AI model's output. We will also measure the augmented user input with the generative AI model's output. Let's first define a cosine similarity function:

```
from sklearn.feature_extraction.text import TfidfVectorizer
from sklearn.metrics.pairwise import cosine_similarity

def calculate_cosine_similarity(text1, text2):
    vectorizer = TfidfVectorizer()
    tfidf = vectorizer.fit_transform([text1, text2])
    similarity = cosine_similarity(tfidf[0:1], tfidf[1:2])
    return similarity[0][0]
```

Then, let's calculate a score that measures the similarity between the user prompt and GPT-4's response:

```
similarity_score = calculate_cosine_similarity(user_prompt, gpt4_response)

print(f"Cosine Similarity Score: {similarity_score:.3f}")
```

The score is low, although the output seemed acceptable for a human:

```
Cosine Similarity Score: 0.396
```

It seems that either we missed something or need to use another metric.

Let's try to calculate the similarity between the augmented input and GPT-4's response:

```
# Example usage with your existing functions
similarity_score = calculate_cosine_similarity(augmented_input, gpt4_response)

print(f"Cosine Similarity Score: {similarity_score:.3f}")
```

The score seems better:

```
Cosine Similarity Score: 0.857
```

Can we use another method? Cosine similarity, when using **Term Frequency-Inverse Document Frequency (TF-IDF)**, relies heavily on exact vocabulary overlap and takes into account important language features, such as semantic meanings, synonyms, or contextual usage. As such, this method may produce lower similarity scores for texts that are conceptually similar but differ in word choice.

In contrast, using Sentence Transformers to calculate similarity involves embeddings that capture deeper semantic relationships between words and phrases. This approach is more effective in recognizing the contextual and conceptual similarity between texts. Let's try this approach.

First, let's install `sentence-transformers`:

```
!pip install sentence-transformers
```

Be careful installing this library at the end of the session, since it may induce potential conflicts with the RAG pipeline's requirements. Depending on a project's needs, this code could be yet another separate pipeline component.

 As of August 2024, using a Hugging Face token is optional. If Hugging Face requires a token, sign up to Hugging Face to obtain an API token, check the conditions, and set up the key as instructed.

We will now use a MiniLM architecture to perform the task with `all-MiniLM-L6-v2`. This model is available through the Hugging Face Model Hub we are using. It's part of the `sentence-transformers` library, which is an extension of the Hugging Face Transformers library. We are using this architecture because it offers a compact and efficient model, with a strong performance in generating meaningful sentence embeddings quickly. Let's now implement it with the following function:

```
from sentence_transformers import SentenceTransformer
model = SentenceTransformer('all-MiniLM-L6-v2')

def calculate_cosine_similarity_with_embeddings(text1, text2):
    embeddings1 = model.encode(text1)
    embeddings2 = model.encode(text2)
    similarity = cosine_similarity([embeddings1], [embeddings2])
    return similarity[0][0]
```

We can now call the function to calculate the similarity between the augmented user input and GPT-4's response:

```
similarity_score = calculate_cosine_similarity_with_embeddings(augmented_input,
gpt4_response)
print(f"Cosine Similarity Score: {similarity_score:.3f}")
```

The output shows that the Sentence Transformer captures semantic similarities between the texts more effectively, resulting in a high cosine similarity score:

```
Cosine Similarity Score: 0.739
```

The choice of metrics depends on the specific requirements of each project phase. *Chapter 3, Building Index-Based RAG with LlamaIndex, Deep Lake, and OpenAI*, will provide advanced metrics when we implement index-based RAG. At this stage, however, the RAG pipeline's three components have been successfully built. Let's summarize our journey and move to the next level!

Summary

In this chapter, we tackled the complexities of using RAG-driven generative AI, focusing on the essential role of document embeddings when handling large datasets. We saw how to go from raw texts to embeddings and store them in vector stores. Vector stores such as Activeloop, unlike parametric generative AI models, provide API tools and visual interfaces that allow us to see embedded text at any moment.

A RAG pipeline detailed the organizational process of integrating OpenAI embeddings into Activeloop Deep Lake vector stores. The RAG pipeline was broken down into distinct components that can vary from one project to another. This separation allows multiple teams to work simultaneously without dependency, accelerating development and facilitating specialized focus on individual aspects, such as data collection, embedding processing, and query generation for the augmented generation AI process.

We then built a three-component RAG pipeline, beginning by highlighting the necessity of specific cross-platform packages and careful system architecture planning. The resources involved were Python functions built from scratch, Activeloop Deep Lake to organize and store the embeddings in a dataset in a vector store, an OpenAI embedding model, and OpenAI's GPT-4o generative AI model. The program guided us through building a three-part RAG pipeline using Python, with practical steps that involved setting up the environment, handling dependencies, and addressing implementation challenges like data chunking and vector store integration.

This journey provided a robust understanding of embedding documents in vector stores and leveraging them for enhanced generative AI outputs, preparing us to apply these insights to real-world AI applications in well-organized processes and teams within an organization. Vector stores enhance the retrieval of documents that require precision in information retrieval. Indexing takes RAG further and increases the speed and relevance of retrievals. The next chapter will take us a step further by introducing advanced indexing methods to retrieve and augment inputs.

Questions

Answer the following questions with *Yes* or *No*:

1. Do embeddings convert text into high-dimensional vectors for faster retrieval in RAG?
2. Are keyword searches more effective than embeddings in retrieving detailed semantic content?
3. Is it recommended to separate RAG pipelines into independent components?
4. Does the RAG pipeline consist of only two main components?

5. Can Activeloop Deep Lake handle both embedding and vector storage?

6. Is the text-embedding-3-small model from OpenAI used to generate embeddings in this chapter?

7. Are data embeddings visible and directly traceable in an RAG-driven system?

8. Can a RAG pipeline run smoothly without splitting into separate components?

9. Is chunking large texts into smaller parts necessary for embedding and storage?

10. Are cosine similarity metrics used to evaluate the relevance of retrieved information?

References

- OpenAI Ada documentation for embeddings: `https://platform.openai.com/docs/guides/embeddings/embedding-models`
- OpenAI GPT documentation for content generation: `https://platform.openai.com/docs/models/gpt-4-turbo-and-gpt-4`
- Activeloop API documentation: `https://docs.deeplake.ai/en/latest/`
- MiniLM model reference: `https://huggingface.co/sentence-transformers/all-MiniLM-L6-v2`

Further reading

- OpenAI's documentation on embeddings: `https://platform.openai.com/docs/guides/embeddings`
- Activeloop documentation: `https://docs.activeloop.ai/`

Unlock this book's exclusive benefits now

This book comes with additional benefits designed to elevate your learning experience.

Note: Have your purchase invoice ready before you begin. `https://www.packtpub.com/unlock/9781836200918`

3

Building Index-Based RAG with LlamaIndex, Deep Lake, and OpenAI

Indexes increase precision and speed performances, but they offer more than that. Indexes transform retrieval-augmented generative AI by adding a layer of transparency. With an index, the source of a response generated by a RAG model is fully traceable, offering visibility into the precise location and detailed content of the data used. This improvement not only mitigates issues like bias and hallucinations but also addresses concerns around copyright and data integrity.

In this chapter, we'll explore how indexed data allows for greater control over generative AI applications. If the output is unsatisfactory, it's no longer a mystery why, since the index allows us to identify and examine the exact data source of the issue. This capability makes it possible to refine data inputs, tweak system configurations, or switch components, such as vector store software and generative models, to achieve better outcomes.

We will begin the chapter by laying out the architecture of an index-based RAG pipeline that will enhance speed, precision, and traceability. We will show how LlamaIndex, Deep Lake, and OpenAI can be seamlessly integrated without having to create all the necessary functions ourselves. This provides a solid base to start building from. Then, we'll introduce the main indexing types we'll use in our programs, such as vector, tree, list, and keyword indexes. Then, we will build a domain-specific drone technology LLM RAG agent that a user can interact with. Drone technology is expanding to all domains, such as fire detection, traffic information, and sports events; hence, I've decided to use it in our example. The goal of this chapter is to prepare an LLM drone technology dataset that we will enhance with multimodal data in the next chapter. We will also illustrate the key indexing types in code.

By the end of this chapter, you'll be adept at manipulating index-based RAG through vector stores, datasets, and LLMs, and know how to optimize retrieval systems and ensure full traceability. You will discover how our integrated toolkit—combining LlamaIndex, Deep Lake, and OpenAI—not only simplifies technical complexities but also frees your time to develop and hone your analytical skills, enabling you to dive deeper into understanding RAG-driven generative AI.

We'll cover the following topics in this chapter:

- Building a semantic search engine with a LlamaIndex framework and indexing methods
- Populating Deep Lake vector stores
- Integration of LlamaIndex, Deep Lake, and OpenAI
- Score ranking and cosine similarity metrics
- Metadata enhancement for traceability
- Query setup and generation configuration
- Introducing automated document ranking
- Vector, tree, list, and keyword indexing types

Why use index-based RAG?

Index-based search takes advanced RAG-driven generative AI to another level. It increases the speed of retrieval when faced with large volumes of data, taking us from raw chunks of data to organized, indexed nodes that we can trace from the output back to the source of a document and its location.

Let's understand the differences between a vector-based similarity search and an index-based search by analyzing the architecture of an index-based RAG.

Architecture

Index-based search is faster than vector-based search in RAG because it directly accesses relevant data using indices, while vector-based search sequentially compares embeddings across all records. We implemented a vector-based similarity search program in *Chapter 2, RAG Embedding Vector Stores with Deep Lake and OpenAI*, as shown in *Figure 3.1*:

- We collected and prepared data in *Pipeline #1: Data Collection and Preparation*
- We embedded the data and stored the prepared data in a vector store in *Pipeline #2: Embeddings and vector store*
- We then ran retrieval queries and generative AI with *Pipeline #3* to process user input, run retrievals based on vector similarity searches, augment the input, generate a response, and apply performance metrics.

This approach is flexible because it gives you many ways to implement each component, depending on the needs of your project.

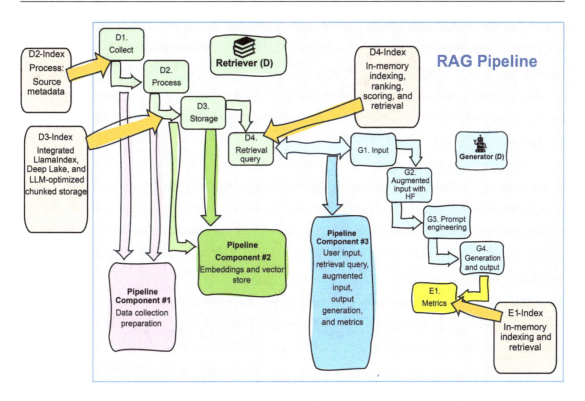

Figure 3.1: RAG-driven generative AI pipelines, as described in Chapter 2, with additional functionality

However, implementing index-based searches will take us into the future of AI, which will be faster, more precise, and traceable. We will follow the same process as in *Chapter 2*, with three pipelines, to make sure that you are ready to work in a team in which the tasks are specialized. Since we are using the same pipelines as in *Chapter 2*, let's add the functions from that chapter to them, as shown in *Figure 3.1*:

- **Pipeline Component #1 and D2-Index:** We will collect data and preprocess it. However, this time, we will prepare the data source one document at a time and store them in separate files. We will then add their name and location to the metadata we load into the vector store. The metadata will help us trace a response all the way back to the exact file that the retrieval function processed. We will have a direct link from a response to the data that it was based on.

- **Pipeline Component #2 and D3-Index:** We will load the data into a vector store by installing and using the innovative integrated `llama-index-vector-stores-deeplake` package, which includes everything we need in an optimized starter scenario: chunking, embedding, storage, and even LLM integration. We have everything we need to get to work on index-based RAG in a few lines of code! This way, once we have a solid program, we can customize and expand the pipelines as we wish, as we did, for example, in *Chapter 2*, when we explicitly chose the LLM models and chunking sizes.

- **Pipeline Component #3 and D4-Index:** We will load the data in a dataset by installing and using the innovative integrated `llama-index-vector-stores-deeplake` package, which includes everything we need to get indexed-based retrieval and generation started, including automated ranking and scoring. The process is seamless and extremely productive. We'll leverage LlamaIndex with Deep Lake to streamline information retrieval and processing. An integrated retriever will efficiently fetch relevant data from the Deep Lake repository, while an LLM agent will then intelligently synthesize and interact with the retrieved information to generate meaningful insights or actions. Indexes are designed for fast retrieval, and we will implement several indexing methods.
- **Pipeline Component #3 and E1-Index:** We will add a time and score metric to evaluate the output.

In the previous chapter, we implemented vector-based similarity search and retrieval. We embedded documents to transform data into high-dimensional vectors. Then, we performed retrieval by calculating distances between vectors. In this chapter, we will go further and create a vector store. However, we will load the data into a dataset that will be reorganized using retrieval indexing types. *Table 3.1* shows the differences between vector-based and index-based search and retrieval methods:

Feature	Vector-based similarity search and retrieval	Index-based vector, tree, list, and keyword search and retrieval
Flexibility	High	Medium (precomputed structure)
Speed	Slower with large datasets	Fast and optimized for quick retrieval
Scalability	Limited by real-time processing	Highly scalable with large datasets
Complexity	Simpler setup	More complex and requires an indexing step
Update Frequency	Easy to update	Requires re-indexing for updates

Table 3.1: Vector-based and index-based characteristics

We will now build a semantic index-based RAG program with Deep Lake, LlamaIndex, and OpenAI.

Building a semantic search engine and generative agent for drone technology

In this section, we will build a semantic index-based search engine and generative AI agent engine using Deep Lake vector stores, LlamaIndex, and OpenAI. As mentioned earlier, drone technology is expanding in domains such as fire detection and traffic control. As such, the program's goal is to provide an index-based RAG agent for drone technology questions and answers. The program will demonstrate how drones use computer vision techniques to identify vehicles and other objects. We will implement the architecture illustrated in *Figure 3.1*, described in the *Architecture* section of this chapter.

 Open `2-Deep_Lake_LlamaIndex_OpenAI_indexing.ipynb` from the GitHub repository of this chapter. The titles of this section are the same as the section titles in the notebook, so you can match the explanations with the code.

We will first begin by installing the environment. Then, we will build the three main pipelines of the program:

- **Pipeline 1**: Collecting and preparing the documents. Using sources like GitHub and Wikipedia, collect and clean documents for indexing.
- **Pipeline 2**: Creating and populating a Deep Lake vector store. Create and populate a Deep Lake vector store with the prepared documents.
- **Pipeline 3**: Index-based RAG for query processing and generation. Applying time and score performances with LLMs and cosine similarity metrics.

When possible, break your project down into separate pipelines so that teams can progress independently and in parallel. The pipelines in this chapter are an example of how this can be done, but there are many other ways to do this, depending on your project. For now, we will begin by installing the environment.

Installing the environment

The environment is mostly the same as in the previous chapter. Let's focus on the packages that integrate LlamaIndex, vector store capabilities for Deep Lake, and also OpenAI modules. This integration is a major step forward to seamless cross-platform implementations:

```
!pip install llama-index-vector-stores-deeplake==0.1.6
```

The program requires additional Deep Lake functionalities:

```
!pip install deeplake==3.9.8
```

The program also requires LlamaIndex functionalities:

```
!pip install llama-index==0.10.64
```

Let's now check if the packages can be properly imported from llama-index, including vector stores for Deep Lake:

```
from llama_index.core import VectorStoreIndex, SimpleDirectoryReader, Document
from llama_index.vector_stores.deeplake import DeepLakeVectorStore
```

With that, we have installed the environment. We will now collect and prepare the documents.

Pipeline 1: Collecting and preparing the documents

In this section, we will collect and prepare the drone-related documents with the metadata necessary to trace the documents back to their source. The goal is to trace a response's content back to the exact chunk of data retrieved to find its source. First, we will create a data directory in which we will load the documents:

```
!mkdir data
```

Now, we will use a heterogeneous corpus for the drone technology data that we will process using BeautifulSoup:

```
import requests
from bs4 import BeautifulSoup
import re
import os

urls = [
    "https://github.com/VisDrone/VisDrone-Dataset",
    "https://paperswithcode.com/dataset/visdrone",
    "https://openaccess.thecvf.com/content_ECCVW_2018/papers/11133/Zhu_
VisDrone-DET2018_The_Vision_Meets_Drone_Object_Detection_in_Image_Challenge_
ECCVW_2018_paper.pdf",
    "https://github.com/VisDrone/VisDrone2018-MOT-toolkit",
    "https://en.wikipedia.org/wiki/Object_detection",
    "https://en.wikipedia.org/wiki/Computer_vision",…
]
```

The corpus contains a list of sites related to drones, computer vision, and related technologies. However, the list also contains noisy links such as `https://keras.io/` and `https://pytorch.org/`, which do *not* contain the specific information we are looking for.

 In real-life projects, we will not always have the luxury of working on perfect, pertinent, structured, and well-formatted data. Our RAG pipelines must be sufficiently robust to retrieve relevant data in a noisy environment.

In this case, we are working with unstructured data in various formats and variable quality as related to drone technology. Of course, in a closed environment, we can work with the persons or organizations that produce the documents, but we must be ready for any type of document in a fast-moving, digital world.

The code will fetch and clean the data, as it did in *Chapter 2*:

```
def clean_text(content):
    # Remove references and unwanted characters
    content = re.sub(r'\[\d+\]', '', content)   # Remove references
    content = re.sub(r'[^\w\s\.]', '', content)  # Remove punctuation (except
periods)
    return content

def fetch_and_clean(url):
    try:
        response = requests.get(url)
```

```
        response.raise_for_status()  # Raise exception for bad responses (e.g.,
404)
        soup = BeautifulSoup(response.content, 'html.parser')

        # Prioritize "mw-parser-output" but fall back to "content" class if not
found
        content = soup.find('div', {'class': 'mw-parser-output'}) or soup.
find('div', {'id': 'content'})
        if content is None:
            return None

        # Remove specific sections, including nested ones
        for section_title in ['References', 'Bibliography', 'External links',
'See also', 'Notes']:
            section = content.find('span', id=section_title)
            while section:
                for sib in section.parent.find_next_siblings():
                    sib.decompose()
                section.parent.decompose()
                section = content.find('span', id=section_title)

        # Extract and clean text
        text = content.get_text(separator=' ', strip=True)
        text = clean_text(text)
        return text
    except requests.exceptions.RequestException as e:
        print(f"Error fetching content from {url}: {e}")
        return None  # Return None on error
```

Each project will require specific names and paths for the original data. In this case, we will introduce an additional function to save each piece of text with the name of its data source, by creating a keyword based on its URL:

```
# Directory to store the output files
output_dir = './data/'
os.makedirs(output_dir, exist_ok=True)

# Processing each URL and writing its content to a separate file
for url in urls:
    article_name = url.split('/')[-1].replace('.html',"")  # Handle .html
extension
    filename = os.path.join(output_dir, article_name + '.txt')  # Create a
filename for the article
```

```
        clean_article_text = fetch_and_clean(url)
    with open(filename, 'w', encoding='utf-8') as file:
        file.write(clean_article_text)


print(f"Content(ones that were possible) written to files in the '{output_dir}'
directory.")
```

The output shows that the goal is achieved, although some documents could not be decoded:

```
WARNING:bs4.dammit:Some characters could not be decoded, and were replaced with
REPLACEMENT CHARACTER.
Content(ones that were possible) written to files in the './data/' directory.
```

Depending on the project's goals, you can choose to investigate and ensure that all documents are retrieved, or estimate that you have enough data for user queries.

If we check ./data/, we will find that each article is now in a separate file, as shown in the content of the directory:

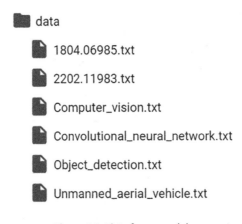

Figure 3.2: List of prepared documents

The program now loads the documents from ./data/:

```
# Load documents
documents = SimpleDirectoryReader("./data/").load_data()
```

The LlamaIndex SimpleDirectoryReader class is designed for working with unstructured data. It recursively scans the directory and identifies and loads all supported file types, such as .txt, .pdf, and .docx. It then extracts the content from each file and returns a list of document objects with its text and metadata, such as the filename and file path. Let's display the first entry of this list of dictionaries of the documents:

```
documents[0]
```

The output shows that the directory reader has provided fully transparent information on the source of its data, including the name of the document, such as 1804.06985.txt in this case:

```
'/content/data/1804.06985.txt', 'file_name': '1804.06985.txt', 'file_type':
'text/plain', 'file_size': 3698, 'creation_date': '2024-05-27', 'last_modified_
date': '2024-05-27'}, excluded_embed_metadata_keys=['file_name', 'file_type',
'file_size', 'creation_date', 'last_modified_date', 'last_accessed_date'],
excluded_llm_metadata_keys=['file_name', 'file_type', 'file_size', 'creation_
date', 'last_modified_date', 'last_accessed_date'], relationships={},
text='High Energy Physics  Theory arXiv1804.06985 hepth Submitted on 19 Apr
2018 Title A Near Horizon Extreme Binary Black Hole Geometry Authors Jacob
Ciafre  Maria J. Rodriguez View a PDF of the paper titled A Near Horizon
Extreme Binary Black Hole Geometry by Jacob Ciafre and Maria J. Rodriguez View
PDF Abstract A new solution of fourdimensional vacuum General Relativity is
presented…
```

The content of this document contains noise that seems unrelated to the drone technology information we are looking for. But that is exactly the point of this program, which aims to do the following:

- Start with all the raw, unstructured, loosely drone-related data we can get our hands on
- Simulate how real-life projects often begin
- Evaluate how well an index-based RAG generative AI program can perform in a challenging environment

Let's now create and populate a Deep Lake vector store in complete transparency.

Pipeline 2: Creating and populating a Deep Lake vector store

In this section, we will create a Deep Lake vector store and populate it with the data in our documents. We will implement a standard tensor configuration with:

- text (str): The text is the content of one of the text files listed in the dictionary of documents. It will be seamless, and chunking will be optimized, breaking the text into meaningful chunks.
- metadata(json): In this case, the metadata will contain the filename source of each chunk of text for full transparency and control. We will see how to access this information in code.
- embedding (float32): The embedding is seamless, using an OpenAI embedding model called directly by the LlamaIndex-Deep Lake-OpenAI package.
- id (str, auto-populated): A unique ID is attributed automatically to each chunk. The vector store will also contain an index, which is a number from 0 to n, but it cannot be used semantically, since it will change each time we modify the dataset. However, the unique ID field will remain unchanged until we decide to optimize it with index-based search strategies, as we will see in the *Pipeline 3: Index-based RAG* section that follows.

The program first defines our vector store and dataset paths:

```
from llama_index.core import StorageContext


vector_store_path = "hub://denis76/drone_v2"
dataset_path = "hub://denis76/drone_v2"
```

Replace the vector store and dataset paths with your account name and the name of the dataset you wish to use:

```
vector_store_path = "hub://[YOUR VECTOR STORE/
```

We then create a vector store, populate it, and create an index over the documents:

```
# overwrite=True will overwrite dataset, False will append it
vector_store = DeepLakeVectorStore(dataset_path=dataset_path, overwrite=True)
storage_context = StorageContext.from_defaults(vector_store=vector_store)
# Create an index over the documents
index = VectorStoreIndex.from_documents(documents, storage_context=storage_
context)
)
```

Notice that `overwrite` is set to `True` to create the vector store and overwrite any existing one. If `overwrite=False`, the dataset will be appended.

The index created will be reorganized by the indexing methods, which will rearrange and create new indexes when necessary. However, the responses will always provide the original source of the data. The output confirms that the dataset has been created and the data is uploaded:

```
Your Deep Lake dataset has been successfully created!
Uploading data to deeplake dataset.
100%|          | 41/41 [00:02<00:00, 18.15it/s]
```

The output also shows the structure of the dataset once it is populated:

```
Dataset(path='hub://denis76/drone_v2', tensors=['text', 'metadata',
'embedding', 'id'])
```

The data is stored in tensors with their type and shape:

tensor	htype	shape	dtype	compression
text	text	(81, 1)	str	None
metadata	json	(81, 1)	str	None
embedding	embedding	(81, 1536)	float32	None
id	text	(81, 1)	str	None

Figure 3.3: Dataset structure

We will now load our dataset in memory:

```
import deeplake
ds = deeplake.load(dataset_path)  # Load the dataset
```

We can visualize the dataset online by clicking on the link provided in the output:

```
/
This dataset can be visualized in Jupyter Notebook by ds.visualize() or at
https://app.activeloop.ai/denis76/drone_v2


hub://denis76/drone_v2 loaded successfully.
This dataset can be visualized in Jupyter Notebook by ds.visualize() or at
https://app.activeloop.ai/denis76/drone_v1
hub://denis76/drone_v2 loaded successfully.
```

We can also decide to add code to display the dataset. We begin by loading the data in a pandas Data-Frame:

```python
import json
import pandas as pd
import numpy as np


# Assuming 'ds' is your loaded Deep Lake dataset

# Create a dictionary to hold the data
data = {}

# Iterate through the tensors in the dataset
for tensor_name in ds.tensors:
    tensor_data = ds[tensor_name].numpy()

    # Check if the tensor is multi-dimensional
    if tensor_data.ndim > 1:
        # Flatten multi-dimensional tensors
        data[tensor_name] = [np.array(e).flatten().tolist() for e in tensor_
data]
    else:
        # Convert 1D tensors directly to lists and decode text
        if tensor_name == "text":
            data[tensor_name] = [t.tobytes().decode('utf-8') if t else "" for t
in tensor_data]
        else:
            data[tensor_name] = tensor_data.tolist()

# Create a Pandas DataFrame from the dictionary
df = pd.DataFrame(data)
```

Then, we create a function to display a record:

```python
# Function to display a selected record
def display_record(record_number):
    record = df.iloc[record_number]
    display_data = {
        "ID": record["id"] if "id" in record else "N/A",
        "Metadata": record["metadata"] if "metadata" in record else "N/A",
        "Text": record["text"] if "text" in record else "N/A",
        "Embedding": record["embedding"] if "embedding" in record else "N/A"
    }
```

Finally, we can select a record and display each field:

```python
# Function call to display a record
rec = 0  # Replace with the desired record number
display_record(rec)
```

The id is a unique string code:

```
ID:
['a89cdb8c-3a85-42ff-9d5f-98f93f414df6']
```

The metadata field contains the information we need to trace the content back to the original file and file path, as well as everything we need to understand this record, from the source to the embedded vector. It also contains the information of the node created from the record's data, which can then be used for the indexing engine we will run in *Pipeline 3*:

- file_path: Path to the file in the dataset (/content/data/1804.06985.txt).
- file_name: Name of the file (`1804.06985.txt`).
- file_type: Type of file (`text/plain`).
- file_size: Size of the file in bytes (`3700`).
- creation_date: Date the file was created (`2024-08-09`).
- last_modified_date: Date the file was last modified (`2024-08-09`).
- _node_content: Detailed content of the node, including the following main items:
    - id_: Unique identifier for the node (`a89cdb8c-3a85-42ff-9d5f-98f93f414df6 `).
    - embedding: Embedding related to the text (null).
    - metadata: Repeated metadata about the file.
    - excluded_embed_metadata_keys: Keys excluded from embedding metadata (not necessary for embedding).
    - excluded_llm_metadata_keys: Keys excluded from LLM metadata (not necessary for an LLM).
    - relationships: Information about relationships to other nodes.

- text: Actual text content of the document. It can be the text itself, an abstract, a summary, or any other approach to optimize search functions.
- start_char_idx: Starting character index of the text.
- end_char_idx: Ending character index of the text.
- text_template: Template for displaying text with metadata.
- metadata_template: Template for displaying metadata.
- metadata_seperator: Separator used in metadata display.
- class_name: Type of node (e.g., `TextNode`).

- _node_type: Type of node (`TextNode`).
- document_id: Identifier for the document (`61e7201d-0359-42b4-9a5f-32c4d67f345e`).
- doc_id: Document ID, same as document_id.
- ref_doc_id: Reference document ID, same as document_id.

The text field contains the field of this chunk of data, not the whole original text:

```
['High Energy Physics  Theory arXiv1804.06985 hepth Submitted on 19 Apr 2018
Title A Near Horizon Extreme Binary Black Hole Geometry Authors Jacob Ciafre
Maria J. Rodriguez View a PDF of the paper titled A Near Horizon Extreme
Binary…
```

The Embedding field contains the embedded vector of the text content:

```
[-0.0009671939187683165, 0.010151553899049759, -0.010979819111526012,
-0.003061748342588544, -0.00865076668560505, 0.02144993655383587,
-0.01412297785282135, -0.02674516849219799, -0.008693241514265537,
-0.03383851423859596, 0.011404570192098618, 0.015956487506628036,
-0.013691147789359093, 0.00885606277340412,…]
```

The structure and format of RAG datasets vary from one domain or project to another. However, the following four columns of this dataset provide valuable information on the evolution of AI:

- id: The id is the index we will be using to organize the chunks of text of the text column in the dataset. The chunks will be transformed into *nodes* that can contain the original text, summaries of the original text, and additional information, such as the source of the data used for the output that is stored in the metadata column. We created this index in **Pipeline 2** of this notebook when we created the vector store. However, we can generate indexes in memory on an existing database that contains no indexes, as we will see in *Chapter 4, Multimodal Modular RAG for Drone Technology*.
- metadata: The metadata was generated automatically in **Pipeline 1** when Deep Lake's SimpleDirectoryReader loaded the source documents in a documents object, and also when the vector store was created. In *Chapter 2, RAG Embedding Vector Stores with Deep Lake and OpenAI*, we only had one file source of data. In this chapter, we stored the data in one file for each data source (URL).

- text: The text processed by Deep Lake's vector store creation functionality that we ran in **Pipeline 2** automatically chunked the data, without us having to configure the size of the chunks, as we did in the *Retrieving a batch of prepared documents* section in *Chapter 2*. Once again, the process is seamless. We will see how smart chunking is done in the *Optimized chunking* section of *Pipeline 3: Index-based RAG* in this chapter.
- embedding: The embedding for each chunk of data was generated through an embedding model that we do not have to configure. We could choose an embedding model, as we did in the *Data embedding and storage* section in *Chapter 2*, *RAG Embedding Vector Stores with Deep Lake and OpenAI*. We selected an embedding model and wrote a function. In this program, Deep Lake selects the embedding model and embeds the data, without us having to write a single line of code.

We can see that embedding, chunking, indexing, and other data processing functions are now encapsulated in platforms and frameworks, such as Activeloop Deep Lake, LlamaIndex, OpenAI, LangChain, Hugging Face, Chroma, and many others. Progressively, the initial excitement of generative AI models and RAG will fade, and they will become industrialized, encapsulated, and commonplace components of AI pipelines. AI is evolving, and it might be helpful to facilitate a platform that offers a default configuration based on effective practices. Then, once we have implemented a basic configuration, we can customize and expand the pipelines as necessary for our projects.

We are now ready to run index-based RAG.

Pipeline 3: Index-based RAG

In this section, we will implement an index-based RAG pipeline using LlamaIndex, which uses the data we have prepared and processed with Deep Lake. We will retrieve relevant information from the heterogeneous (noise-containing) drone-related document collection and synthesize the response through OpenAI's LLM models. We will implement four index engines:

- **Vector Store Index Engine:** Creates a vector store index from the documents, enabling efficient similarity-based searches.
- **Tree Index:** Builds a hierarchical tree index from the documents, offering an alternative retrieval structure.
- **List Index:** Constructs a straightforward list index from the documents.
- **Keyword Table Index:** Creates an index based on keywords extracted from the documents.

We will implement querying with an LLM:

- **Query Response and Source:** Queries the index with user input, retrieves the relevant documents, and returns a synthesized response along with source information.

We will measure the responses with a *time-weighted average metric with LLM score and cosine similarity* that calculates a time-weighted average, based on retrieval and similarity scores. The content and execution times might vary from one run to another due to the stochastic algorithms implemented.

User input and query parameters

The user input will be the reference question for the four index engines we will run. We will evaluate each response based on the index engine's retrievals and measure the outputs, using time and score ratios. The input will be submitted to the four index and query engines we will build later.

The user input is:

```
user_input="How do drones identify vehicles?"
```

The four query engines that implement an LLM (in this case, an OpenAI model) will seamlessly be called with the same parameters. The three parameters that we will set are:

```
#similarity_top_k
k=3
#temperature
temp=0.1
#num_output
mt=1024
```

These key parameters are:

- k=3: The query engine will be required to find the top 3 most probable responses by setting the top-k (most probable choices) to 3. In this case, k will serve as a ranking function that will force the LLM to select the top documents.
- temp=0.1: A low temperature such as 0.1 will encourage the LLM to produce precise results. If the temperature is increased to 0.9, for example, the response will be more creative. However, in this case, we are exploring drone technology, which requires precision.
- mt=1024: This parameter will limit the number of tokens of the output to 1,024.

The user input and parameters will be applied to the four query engines. Let's now build the cosine similarity metric.

Cosine similarity metric

The cosine similarity metric was described in the *Evaluating the Output with the Cosine Similarity* section in *Chapter 2*. If necessary, take the time to go through that section again. Here, we will create a function for the responses:

```
from sklearn.feature_extraction.text import TfidfVectorizer
from sklearn.metrics.pairwise import cosine_similarity

from sentence_transformers import SentenceTransformer
model = SentenceTransformer('all-MiniLM-L6-v2')

def calculate_cosine_similarity_with_embeddings(text1, text2):
    embeddings1 = model.encode(text1)
```

```
embeddings2 = model.encode(text2)
similarity = cosine_similarity([embeddings1], [embeddings2])
return similarity[0][0]
```

The function uses `sklearn` and also Hugging Face's `SentenceTransformer`. The program first creates the vector store engine.

Vector store index query engine

`VectorStoreIndex` is a type of index within LlamaIndex that implements vector embeddings to represent and retrieve information from documents. These documents with similar meanings will have embeddings that are closer together in the vector space, as we explored in the previous chapter. However, this time, the `VectorStoreIndex` does not automatically use the existing Deep Lake vector store. It can create a new in-memory vector index, re-embed the documents, and create a new index structure. We will take this approach further in *Chapter 4, Multimodal Modular RAG for Drone Technology*, when we implement a dataset that contains no indexes or embeddings.

 There is no silver bullet to deciding which indexing method is suitable for your project! The best way to make a choice is to test the vector, tree, list, and keyword indexes introduced in this chapter.

We will first create the vector store index:

```
from llama_index.core import VectorStoreIndex
vector_store_index = VectorStoreIndex.from_documents(documents)
```

We then display the vector store index we created:

```
print(type(vector_store_index))
```

We will receive the following output, which confirms that the engine was created:

```
<class 'llama_index.core.indices.vector_store.base.VectorStoreIndex'>
```

We now need a query engine to retrieve and synthesize the document(s) retrieved with an LLM—in our case, an OpenAI model (installed with `!pip install llama-index-vector-stores-deeplake==0.1.2`):

```
vector_query_engine = vector_store_index.as_query_engine(similarity_top_k=k,
temperature=temp, num_output=mt)
```

We defined the parameters of the query engine in the *User input and query parameters* subsection. We can now query the dataset and generate a response.

Query response and source

Let's define a function that will manage the query and return information on the content of the response:

```python
import pandas as pd
import textwrap

def index_query(input_query):
    response = vector_query_engine.query(input_query)

    # Optional: Print a formatted view of the response (remove if you don't
need it in the output)
    print(textwrap.fill(str(response), 100))

    node_data = []
    for node_with_score in response.source_nodes:
        node = node_with_score.node
        node_info = {
            'Node ID': node.id_,
            'Score': node_with_score.score,
            'Text': node.text
        }
        node_data.append(node_info)

    df = pd.DataFrame(node_data)

    # Instead of printing, return the DataFrame and the response object
    return df, response,
```

index_query(input_query) executes a query using a vector query engine and processes the results into a structured format. The function takes an input query and retrieves relevant information, using the query engine in a pandas DataFrame: Node ID, Score, File Path, Filename, and Text.

The code will now call the query:

```python
import time
#start the timer
start_time = time.time()
df, response = index_query(user_input)
# Stop the timer
end_time = time.time()
```

```
# Calculate and print the execution time
elapsed_time = end_time - start_time
print(f"Query execution time: {elapsed_time:.4f} seconds")


print(df.to_markdown(index=False, numalign="left", stralign="left"))  # Display
the DataFrame using markdown
```

We will evaluate the time it takes for the query to retrieve the relevant data and generate a response synthesis with the LLM (in this case, an OpenAI model). The output of the semantic search first returns a response synthesized by the LLM:

```
Drones can automatically identify vehicles across different cameras with
different viewpoints and hardware specifications using reidentification
methods.
```

The output then displays the elapsed time of the query:

```
Query execution time: 0.8831 seconds
```

The output now displays node information. The score of each node of three k=3 documents was retrieved with their text excerpts:

```
| Node ID                              | Score    | Text
|:-------------------------------------|:---------|:-------------------------------------------
| 4befdb13-305d-42db-a616-5d9932c17ac8 | 0.833274 | ['These activities can be carried out with
| bdefbbba-1b1a-4aea-812c-a2f6acb4dfb9 | 0.828414 | ['Degree of autonomy  edit  Drones could
| 381255c4-3bdb-455a-be73-d99cc6b46537 | 0.826526 | ['UAVs with generally nonlethal payloads
```

Figure 3.4: Node information output

The ID of the node guarantees full transparency and can be traced back to the original document, even when the index engines re-index the dataset. We can obtain the node source of the first node, for example, with the following code:

```
nodeid=response.source_nodes[0].node_id
nodeid
```

The output provides the node ID:

```
4befdb13-305d-42db-a616-5d9932c17ac8
```

We can drill down and retrieve the full text of the node containing the document that was synthesized by the LLM:

```
response.source_nodes[0].get_text()
```

The output will display the following text:

```
['These activities can be carried out with different approaches that include
photogrammetry SfM thermography multispectral images 3D field scanning NDVI
maps etc. Agriculture forestry and environmental studies edit Main article
Agricultural drone As global demand for food production grows exponentially
resources are depleted farmland is…
```

We can also peek into the nodes and retrieve their chunk size.

Optimized chunking

We can predefine the chunk size, or we can let LlamaIndex select it for us. In this case, the code determines the chunk size automatically:

```
for node_with_score in response.source_nodes:
    node = node_with_score.node  # Extract the Node object from NodeWithScore

    chunk_size = len(node.text)
    print(f"Node ID: {node.id_}, Chunk Size: {chunk_size} characters")
```

The advantage of an automated chunk size is that it can be variable. For example, in this case, the chunk size shown in the size of the output nodes is probably in the 4000-to-5500-character range:

```
Node ID: 83a135c6-dddd-402e-9423-d282e6524160, Chunk Size: 4417 characters
Node ID: 7b7b55fe-0354-45bc-98da-0a715ceaaab0, Chunk Size: 1806 characters
Node ID: 18528a16-ce77-46a9-bbc6-5e8f05418d95, Chunk Size: 3258 characters
```

The chunking function does not linearly cut content but optimizes the chunks for semantic search.

Performance metric

We will also implement a performance metric based on the accuracy of the queries and the time elapsed. This function calculates and prints a performance metric for a query, along with its execution time. The metric is based on the weighted average relevance scores of the retrieved information, divided by the time it took to get the results. Higher scores indicate better performance.

We first calculate the sum of the scores and the average score, and then we divide the weighted average by the time elapsed to perform the query:

```python
import numpy as np

def info_metrics(response):
  # Calculate the performance (handling None scores)
  scores = [node.score for node in response.source_nodes if node.score is not
None]
  if scores:  # Check if there are any valid scores
      weights = np.exp(scores) / np.sum(np.exp(scores))
      perf = np.average(scores, weights=weights) / elapsed_time
  else:
      perf = 0  # Or some other default value if all scores are None
```

The result is a ratio based on the average weight divided by the elapsed time:

```python
perf = np.average(scores, weights=weights) / elapsed_time
```

We can then call the function:

```python
info_metrics(response)
```

The output provides an estimation of the quality of the response:

```
Average score: 0.8374
Query execution time: 1.3266 seconds
Performance metric: 0.6312
```

This performance metric is not an absolute value. It's an indicator that we can use to compare this output with the other index engines. It may also vary from one run to another, due to the stochastic nature of machine learning algorithms. Additionally, the quality of the output depends on the user's subjective perception. In any case, this metric will help compare the query engines' performances in this chapter.

We can already see that the average score is satisfactory, even though we loaded heterogeneous and sometimes unrelated documents in the dataset. The integrated retriever and synthesizer functionality of LlamaIndex, Deep Lake, and OpenAI have proven to be highly effective.

Tree index query engine

The tree index in LlamaIndex creates a hierarchical structure for managing and querying text documents efficiently. However, think of something other than a classical hierarchical structure! The tree index engine optimizes the hierarchy, content, and order of the nodes, as shown in *Figure 3.5*:

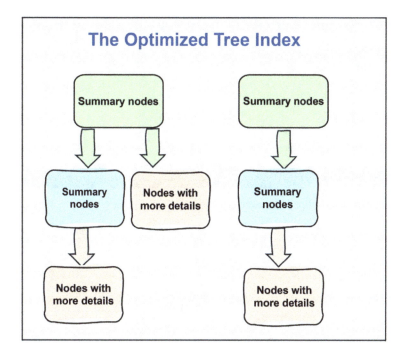

Figure 3.5: Optimized tree index

The tree index organizes documents in a tree structure, with broader summaries at higher levels and detailed information at lower levels. Each node in the tree summarizes the text it covers. The tree index is efficient for large datasets and queries large collections of documents rapidly by breaking them down into manageable optimized chunks. Thus, the optimization of the tree structure allows for rapid retrieval by traversing the relevant nodes without wasting time.

Organizing this part of the pipeline and adjusting parameters such as tree depth and summary methods can be a specialized task for a team member. Depending on the project and workload, working on the tree structure could be part of **Pipeline 2** when creating and populating a vector store. Alternatively, the tree structure can be created in memory at the beginning of each session. The flexibility of the structure and implementation of tree structures and index engines, in general, can be a fascinating and valuable specialization in a RAG-driven generative AI team.

In this index model, the LLM (an OpenAI model in this case) acts like it is answering a multiple-choice question when selecting the best nodes during a query. It analyzes the query, compares it with the summaries of the current node's children, and decides which path to follow to find the most relevant information.

The integrated LlamaIndex-Deep Lake-OpenAI process in this chapter is industrializing components seamlessly, taking AI to another level. LLM models can now be used for embedding, document ranking, and conversational agents. The market offers various language models from providers like OpenAI, Cohere, AI21 Labs, and Hugging Face. LLMs have evolved from the early days of being perceived as magic to becoming industrialized, seamless, multifunctional, and integrated components of broader AI pipelines.

Let's create a tree index in two lines of code:

```
from llama_index.core import TreeIndex
tree_index = TreeIndex.from_documents(documents)
```

The code then checks the class we just created:

```
print(type(tree_index))
```

The output confirms that we are in the `TreeIndex` class:

```
<class 'llama_index.core.indices.tree.base.TreeIndex'>
```

We can now make our tree index the query engine:

```
tree_query_engine = tree_index.as_query_engine(similarity_top_k=k,
temperature=temp, num_output=mt)
```

The parameters of the LLM are those defined in the *User input and query parameters* section. The code now calls the query, measures the time elapsed, and processes the response:

```
import time
import textwrap
# Start the timer
start_time = time.time()
response = tree_query_engine.query(user_input)
# Stop the timer
end_time = time.time()
# Calculate and print the execution time
elapsed_time = end_time - start_time
print(f"Query execution time: {elapsed_time:.4f} seconds")

print(textwrap.fill(str(response), 100))
```

The query time and the response are both satisfactory:

```
Query execution time: 4.3360 seconds
Drones identify vehicles using computer vision technology related to object
detection. This
technology involves detecting instances of semantic objects of a certain class,
such as vehicles, in
digital images and videos. Drones can be equipped with object detection
algorithms, such as YOLOv3
models trained on datasets like COCO, to detect vehicles in real-time by
analyzing the visual data
captured by the drone's cameras.
```

Let's apply a performance metric to the output.

Performance metric

This performance metric will calculate the cosine similarity defined in the *Cosine similarity metric* section between the user input and the response of our RAG pipeline:

```
similarity_score = calculate_cosine_similarity_with_embeddings(user_input,
str(response))
print(f"Cosine Similarity Score: {similarity_score:.3f}")
print(f"Query execution time: {elapsed_time:.4f} seconds")
performance=similarity_score/elapsed_time
print(f"Performance metric: {performance:.4f}")
```

The output shows that although the quality of the response was satisfactory, the execution time was slow, which brings the performance metric down:

```
Cosine Similarity Score: 0.731
Query execution time: 4.3360 seconds
Performance metric: 0.1686
```

Of course, the execution time depends on the server (power) and the data (noise). As established earlier, the execution times might vary from one run to another, due to the stochastic algorithms used. Also, when the dataset increases in volume, the execution times of all the indexing types may change.

The list index query engine may or may not be better in this case. Let's run it to find out.

List index query engine

Don't think of ListIndex as simply a list of nodes. The query engine will process the user input and each document as a prompt for an LLM. The LLM will evaluate the semantic similarity relationship between the documents and the query, thus implicitly ranking and selecting the most relevant nodes. LlamaIndex will filter the documents based on the rankings obtained, and it can also take the task further by synthesizing information from multiple nodes and documents.

We can see that the selection process with an LLM is not rule-based. Nothing is predefined, which means that the selection is prompt-based by combining the user input with a collection of documents. The LLM evaluates each document in the list *independently*, assigning a score based on its perceived relevance to the query. This score isn't relative to other documents; it's a measure of how well the LLM thinks the current document answers the question. Then, the top-k documents are retained by the query engine if we wish, as in the function used in this section.

Like the tree index, the list index can also be created in two lines of code:

```
from llama_index.core import ListIndex
list_index = ListIndex.from_documents(documents)
```

The code verifies the class that we are using:

```
print(type(list_index))
```

The output confirms that we are in the `list` class:

```
<class 'llama_index.core.indices.list.base.SummaryIndex'>
```

The list index is a `SummaryIndex`, which shows the large amount of document summary optimization that is running under the hood! We can now utilize our list index as a query engine in the seamless framework provided by LlamaIndex:

```
list_query_engine = list_index.as_query_engine(similarity_top_k=k,
temperature=temp, num_output=mt)
```

The LLM parameters remain unchanged so that we can compare the indexing types. We can now run our query, wrap the response up, and display the output:

```
#start the timer
start_time = time.time()
response = list_query_engine.query(user_input)
# Stop the timer
end_time = time.time()
# Calculate and print the execution time
elapsed_time = end_time - start_time
print(f"Query execution time: {elapsed_time:.4f} seconds")

print(textwrap.fill(str(response), 100))
```

The output shows a longer execution time but an acceptable response:

```
Query execution time: 16.3123 seconds
Drones can identify vehicles through computer vision systems that process image
data captured by
cameras mounted on the drones. These systems use techniques like object
recognition and detection to
analyze the images and identify specific objects, such as vehicles, based on
predefined models or
features. By processing the visual data in real-time, drones can effectively
identify vehicles in
their surroundings.
```

The execution time is longer because the query goes through a list, not an optimized tree. However, we cannot draw conclusions from this because each project or even each sub-task of a project has different requirements. Next, let's apply the performance metric.

Performance metric

We will use the cosine similarity, as we did for the tree index, to evaluate the similarity score:

```
similarity_score = calculate_cosine_similarity_with_embeddings(user_input,
str(response))
```

```
print(f"Cosine Similarity Score: {similarity_score:.3f}")
print(f"Query execution time: {elapsed_time:.4f} seconds")
performance=similarity_score/elapsed_time
print(f"Performance metric: {performance:.4f}")
```

The performance metric is lower than the tree index due to the longer execution time:

```
Cosine Similarity Score: 0.775
Query execution time: 16.3123 seconds
Performance metric: 0.0475
```

Again, remember that this execution time may vary from one run to another, due to the stochastic algorithms implemented.

If we look back at the performance metric of each indexing type, we can see that, for the moment, the vector store index was the fastest. Once again, let's not jump to conclusions. Each project might produce surprising results, depending on the type and complexity of the data processed. Next, let's examine the keyword index.

Keyword index query engine

KeywordTableIndex is a type of index in LlamaIndex, designed to extract keywords from your documents and organize them in a table-like structure. This structure makes it easier to query and retrieve relevant information based on specific keywords or topics. Once again, don't think about this function as a simple list of extracted keywords. The extracted keywords are organized into a table-like format where each keyword is associated with an ID that points to the related nodes.

The program creates the keyword index in two lines of code:

```
from llama_index.core import KeywordTableIndex
keyword_index = KeywordTableIndex.from_documents(documents)
```

Let's extract the data and create a pandas DataFrame to see how the index is structured:

```
# Extract data for DataFrame
data = []
for keyword, doc_ids in keyword_index.index_struct.table.items():
    for doc_id in doc_ids:
        data.append({"Keyword": keyword, "Document ID": doc_id})

# Create the DataFrame
df = pd.DataFrame(data)
df
```

The output shows that each keyword is associated with an ID that contains a document or a summary, depending on the way LlamaIndex optimizes the index:

	Keyword	Document ID
0	black	48696b4b-978e-46b1-a5b3-f451d5265f42
1	asymptotically flat	48696b4b-978e-46b1-a5b3-f451d5265f42
2	high energy physics	48696b4b-978e-46b1-a5b3-f451d5265f42
3	entropy	48696b4b-978e-46b1-a5b3-f451d5265f42
4	extreme binary black hole geometry	48696b4b-978e-46b1-a5b3-f451d5265f42
...	...	...

Figure 3.6: Keywords linked to document IDs in a DataFrame

We now define the keyword index as the query engine:

```
keyword_query_engine = keyword_index.as_query_engine(similarity_top_k=k,
temperature=temp, num_output=mt)
```

Let's run the keyword query and see how well and fast it can produce a response:

```
import time

# Start the timer
start_time = time.time()

# Execute the query (using .query() method)
response = keyword_query_engine.query(user_input)

# Stop the timer
end_time = time.time()

# Calculate and print the execution time
elapsed_time = end_time - start_time
print(f"Query execution time: {elapsed_time:.4f} seconds")

print(textwrap.fill(str(response), 100))
```

The output is satisfactory, as well as the execution time:

```
Query execution time: 2.4282 seconds
Drones can identify vehicles through various means such as visual recognition
using onboard cameras, sensors, and image processing algorithms. They can
also utilize technologies like artificial intelligence and machine learning
to analyze and classify vehicles based on their shapes, sizes, and movement
patterns. Additionally, drones can be equipped with specialized software for
object detection and tracking to identify vehicles accurately.
```

We can now measure the output with a performance metric.

Performance metric

The code runs the same metric as for the tree and list index:

```
similarity_score = calculate_cosine_similarity_with_embeddings(user_input,
str(response))
print(f"Cosine Similarity Score: {similarity_score:.3f}")
print(f"Query execution time: {elapsed_time:.4f} seconds")
performance=similarity_score/elapsed_time
print(f"Performance metric: {performance:.4f}")
```

The performance metric is acceptable:

```
Cosine Similarity Score: 0.801
Query execution time: 2.4282 seconds
Performance metric: 0.3299
```

Once again, we can draw no conclusions. The results of all the indexing types are relatively satisfactory. However, each project comes with its dataset complexity and machine power availability. Also, the execution times may vary from one run to another, due to the stochastic algorithms employed.

With that, we have reviewed some of the main indexing types and retrieval strategies. Let's summarize the chapter and move on to multimodal modular retrieval and generation strategies.

Summary

This chapter explored the transformative impact of index-based search on RAG and introduced a pivotal advancement: *full traceability*. The documents become nodes that contain chunks of data, with the source of a query leading us all the way back to the original data. Indexes also increase the speed of retrievals, which is critical as the volume of datasets increases. Another pivotal advance is the integration of technologies such as LlamaIndex, Deep Lake, and OpenAI, which are emerging in another era of AI. The most advanced AI models, such as OpenAI GPT-4o, Hugging Face, and Cohere, are becoming seamless *components* in a RAG-driven generative AI pipeline, like GPUs in a computer.

We started by detailing the architecture of an index-based RAG generative AI pipeline, illustrating how these sophisticated technologies can be seamlessly integrated to boost the creation of advanced indexing and retrieval systems. The complexity of AI implementation is changing the way we organize separate pipelines and functionality for a team working in parallel on projects that scale and involve large amounts of data. We saw how every response generated can be traced back to its source, providing clear visibility into the origins and accuracy of the information used. We illustrated the advanced RAG technology implemented through drone technology.

Throughout the chapter, we introduced the essential tools to build these systems, including vector stores, datasets, chunking, embedding, node creation, ranking, and indexing methods. We implemented the LlamaIndex framework, Deep Lake vector stores, and OpenAI's models. We also built a Python program that collects data and adds critical metadata to pinpoint the origin of every chunk of data in a dataset. We highlighted the pivotal role of indexes (vector, tree, list, and keyword types) in giving us greater control over generative AI applications, enabling precise adjustments and improvements.

We then thoroughly examined indexed-based RAG through detailed walkthroughs in Python notebooks, guiding you through setting up vector stores, conducting advanced queries, and ensuring the traceability of AI-generated responses. We introduced metrics based on the quality of a response and the time elapsed to obtain it. Exploring drone technology with LLMs showed us the new skillsets required to build solid AI pipelines, and we learned how drone technology involves computer vision and, thus, multimodal nodes.

In the upcoming chapter, we include multimodal data in our datasets and expand multimodular RAG.

Questions

Answer the following questions with *Yes* or *No*:

- Do indexes increase precision and speed in retrieval-augmented generative AI?
- Can indexes offer traceability for RAG outputs?
- Is index-based search slower than vector-based search for large datasets?
- Does LlamaIndex integrate seamlessly with Deep Lake and OpenAI?
- Are tree, list, vector, and keyword indexes the only types of indexes?
- Does the keyword index rely on semantic understanding to retrieve data?
- Is LlamaIndex capable of automatically handling chunking and embedding?
- Are metadata enhancements crucial for ensuring the traceability of RAG-generated outputs?
- Can real-time updates easily be applied to an index-based search system?
- Is cosine similarity a metric used in this chapter to evaluate query accuracy?

References

- LlamaIndex: https://docs.llamaindex.ai/en/stable/
- Activeloop Deep Lake: https://docs.activeloop.ai/
- OpenAI: https://platform.openai.com/docs/overview

Further reading

- High-Level Concepts (RAG), LlamaIndex: https://docs.llamaindex.ai/en/stable/getting_started/concepts/

Learn further in a live workshop with the author

If you are curious how RAG-based systems scale into agentic workflows using context engineering, consider registering for the author's live workshop on 24 January 2026, which extends the ideas discussed here into hands-on practice.

https://packt.link/gk5Yu

4

Multimodal Modular RAG for Drone Technology

We will take generative AI to the next level with modular RAG in this chapter. We will build a system that uses different components or modules to handle different types of data and tasks. For example, one module processes textual information using LLMs, as we have done until the last chapter, while another module manages image data, identifying and labeling objects within images. Imagine using this technology in drones, which have become crucial across various industries, offering enhanced capabilities for aerial photography, efficient agricultural monitoring, and effective search and rescue operations. They even use advanced computer vision technology and algorithms to analyze images and identify objects like pedestrians, cars, trucks, and more. We can then activate an LLM agent to retrieve, augment, and respond to a user's question.

In this chapter, we will build a multimodal modular RAG program to generate responses to queries about drone technology using text and image data from multiple sources. We will first define the main aspects of modular RAG, multimodal data, multisource retrieval, modular generation, and augmented output. We will then build a multimodal modular RAG-driven generative AI system in Python applied to drone technology with LlamaIndex, Deep Lake, and OpenAI.

Our system will use two datasets: the first one containing textual information about drones that we built in the previous chapter and the second one containing drone images and labels from Activeloop. We will use Deep Lake to work with multimodal data, LlamaIndex for indexing and retrieval, and generative queries with OpenAI LLMs. We will add multimodal augmented outputs with text and images. Finally, we will build performance metrics for the text responses and introduce an image recognition metric with GPT-4o, OpenAI's powerful **Multimodal LLM (MMLLM)**. By the end of the chapter, you will know how to build a multimodal modular RAG workflow leveraging innovative multimodal and multisource functionalities.

This chapter covers the following topics:

- Multimodal modular RAG
- Multisource retrieval
- OpenAI LLM-guided multimodal multisource retrieval
- Deep Lake multimodal datasets
- Image metadata-based retrieval
- Augmented multimodal output

Let's begin by defining multimodal modular RAG.

What is multimodal modular RAG?

Multimodal data combines different forms of information, such as text, images, audio, and video, to enrich data analysis and interpretation. Meanwhile, a system is a modular RAG system when it utilizes distinct modules for handling different data types and tasks. Each module is specialized; for example, one module will focus on text and another on images, demonstrating a sophisticated integration capability that enhances response generation with retrieved multimodal data.

The program in this chapter will also be multisource through the two datasets we will use. We will use the LLM dataset on the drone technology built in the previous chapter. We will also use the Deep Lake multimodal VisDrone dataset, which contains thousands of labeled images captured by drones.

We have selected drones for our example since drones have become crucial across various industries, offering enhanced capabilities for aerial photography, efficient agricultural monitoring, and effective search and rescue operations. They also facilitate wildlife tracking, streamline commercial deliveries, and enable safer infrastructure inspections. Additionally, drones support environmental research, traffic management, and firefighting. They can enhance surveillance for law enforcement, revolutionizing multiple fields by improving accessibility, safety, and cost-efficiency.

Figure 4.1 contains the workflow we will implement in this chapter. It is based on the generative RAG ecosystem illustrated in *Figure 1.3* from *Chapter 1, Why Retrieval-Augmented Generation?*. We added embedding and indexing functionality in the previous chapters, but this chapter will focus on retrieval and generation. The system we will build blurs the lines between retrieval and generation since the generator is intensively used for retrieving (seamless scoring and ranking) as well as generating in the chapter's notebook.

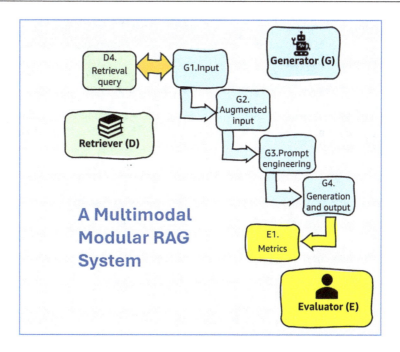

Figure 4.1: A multimodal modular RAG system

This chapter aims to build an educational modular RAG question-answering system focused on drone technology. You can rely on the functionality implemented in the notebooks of the preceding chapters, such as Deep Lake for vectors in *Chapter 2, RAG Embedding Vector Stores with Deep Lake and OpenAI*, and indices with LlamaIndex in *Chapter 3, Building Index-based RAG with LlamaIndex, Deep Lake, and OpenAI*. If necessary, take your time to go back to the previous chapters and have a look.

Let's go through the multimodal, multisource, modular RAG ecosystem in this chapter, represented in *Figure 4.1*. We will use the titles and subsections in this chapter represented in italics. Also, each phase is preceded by its location in *Figure 4.1*.

- **(D4)** *Loading the LLM dataset* created in *Chapter 3*, which contains textual data on drones.
- **(D4)** *Initializing the LLM query engine* with a LlamaIndex vector store index using `VectorStoreIndex` and setting the created index for the query engine, which overlaps with **(G4)** as both a retriever and a generator with the OpenAI GPT model.
- **(G1)** Defining the *user input for multimodal modular RAG* for both the LLM query engine (for the textual dataset) and the multimodal query engine (for the `VisDrone` dataset).

  Once the textual dataset has been loaded, the query engine has been created, and the user input has been defined as a baseline query for the textual dataset and the multimodal dataset, the process continues by generating a response for the textual dataset created in *Chapter 2*.

- While *querying the textual dataset*, **(G1)**, **(G2)**, and **(G4)** overlap in the same seamless LlamaIndex process that retrieves data and generates content. The response is saved as `llm_response` for the duration of the session.

Now, the multimodal `VisDrone` dataset will be loaded into memory and queried:

- **(D4)** The multimodal process begins by *loading and visualizing the multimodal dataset*. The program then continues by *navigating the multimodal dataset structure, selecting an image*, and *adding bounding boxes*.

The same process as for the textual dataset is then applied to the `VisDrone` multimodal dataset:

- **(D4)** *Building a multimodal query engine* with LlamaIndex by creating a vector store index based on `VisDrone` data using `VectorStoreIndex` and setting the created index for the query engine, which overlaps with **(G4)** as both a retriever and a generator with OpenAI GPT.
- **(G1)** The user input for the multimodal search engine is the same as the *user input for multimodal modular RAG* since it is used for both the LLM query engine (for the textual dataset) and the multimodal query engine (for the `VisDrone` dataset).

The multimodal `VisDrone` dataset will now be loaded and indexed, and the query engine is ready. The purpose of **(G1)** user input is for the LlamaIndex query engine to retrieve relevant documents from VisDrone using an LLM—in this case, an OpenAI model. Then, the retrieval functions will trace the response back to its source in the multimodal dataset to find the image of the source nodes. We are, in fact, using the query engine to reach an image through its textual response:

- **(G1)**, **(G2)**, and **(G4)** overlap in a seamless LlamaIndex query when running a query on the `VisDrone` multimodal dataset.
- Processing the response **(G4)** to find the source node and retrieve its image leads us back to **(D4)** for image retrieval. This leads to selecting and processing the image of the source node.

At this point, we now have the textual and the image response. We can then build a summary and apply an accuracy performance metric after having visualized the time elapsed for each phase as we built the program:

- **(G4)** We present a merged output with the LLM response and the augmented output with the image of the multimodal response in a *multimodal modular summary*.
- **(E)** Finally, we create an *LLM performance metric* and a *multimodal performance metric*. We then sum them up as a *multimodal modular RAG performance metric*.

We can draw two conclusions from this multimodal modular RAG system:

- The system we are building in this chapter is one of the many ways RAG-driven generative AI can be designed in real-life projects. Each project will have its specific needs and architecture.
- The rapid evolution from generative AI to the complexity of RAG-driven generative AI requires the corresponding development of seamlessly integrated cross-platform components such as LlamaIndex, Deep Lake, and OpenAI in this chapter. These platforms are also integrated with many other frameworks, such as Pinecone and LangChain, which we will discuss in *Chapter 6, Scaling RAG Bank Customer Data with Pinecone*.

Now, let's dive into Python and build the multimodal modular RAG program.

Building a multimodal modular RAG program for drone technology

In the following sections, we will build a multimodal modular RAG-driven generative system from scratch in Python, step by step. We will implement:

- LlamaIndex-managed OpenAI LLMs to process and understand text about drones
- Deep Lake multimodal datasets containing images and labels of drone images taken
- Functions to display images and identify objects within them using bounding boxes
- A system that can answer questions about drone technology using both text and images
- Performance metrics aimed at measuring the accuracy of the modular multimodal responses, including image analysis with GPT-4o

Also, make sure you have created the LLM dataset in *Chapter 2* since we will be loading it in this section. However, you can read this chapter without running the notebook since it is self-contained with code and explanations. Now, let's get to work!

Open the `Multimodal_Modular_RAG_Drones.ipynb` notebook in the GitHub repository for this chapter at `https://github.com/Denis2054/RAG-Driven-Generative-AI/tree/main/Chapter04`. The packages installed are the same as those listed in the *Installing the environment* section of the previous chapter. Each of the following sections will guide you through building the multimodal modular notebook, starting with the LLM module. Let's go through each section of the notebook step by step.

Loading the LLM dataset

We will load the drone dataset created in *Chapter 3*. Make sure to insert the path to your dataset:

```
import deeplake
dataset_path_llm = "hub://denis76/drone_v2"
ds_llm = deeplake.load(dataset_path_llm)
```

The output will confirm that the dataset is loaded and will display the link to your dataset:

```
This dataset can be visualized in Jupyter Notebook by ds.visualize() or at
https://app.activeloop.ai/denis76/drone_v2
hub://denis76/drone_v2 loaded successfully.
```

The program now creates a dictionary to hold the data to load it into a pandas DataFrame to visualize it:

```
import json
import pandas as pd
import numpy as np

# Create a dictionary to hold the data
data_llm = {}
```

```
# Iterate through the tensors in the dataset
for tensor_name in ds_llm.tensors:
    tensor_data = ds_llm[tensor_name].numpy()

    # Check if the tensor is multi-dimensional
    if tensor_data.ndim > 1:
        # Flatten multi-dimensional tensors
        data_llm[tensor_name] = [np.array(e).flatten().tolist() for e in
tensor_data]
    else:
        # Convert 1D tensors directly to lists and decode text
        if tensor_name == "text":
            data_llm[tensor_name] = [t.tobytes().decode('utf-8') if t else ""
for t in tensor_data]
        else:
            data_llm[tensor_name] = tensor_data.tolist()

# Create a Pandas DataFrame from the dictionary
df_llm = pd.DataFrame(data_llm)
df_llm
```

💡 **Quick tip:** Enhance your coding experience with the **AI Code Explainer** and **Quick Copy** features. Open this book in the next-gen Packt Reader. Click the **Copy** button (**1**) to quickly copy code into your coding environment, or click the **Explain** button (**2**) to get the AI assistant to explain a block of code to you.

```
                                                          Copy      Explain

function calculate(a, b) {                                  1          2
    return {sum: a + b};
};
```

🔖 **The next-gen Packt Reader** is included for free with the purchase of this book. Unlock it by scanning the QR code below or visiting `https://www.packtpub.com/unlock/9781836200918`.

The output shows the text dataset with its structure: embedding (vectors), id (unique string identifier), metadata (in this case, the source of the data), and text, which contains the content:

	embedding	id	metadata	text
0	[-0.0009671939187683165, 0.010151553899049759,...	[a89cdb8c-3a85-42ff-9d5f-98f93f414df6]	[{'file_path': '/content/data/1804.06985.txt',...	[High Energy Physics Theory arXiv1804.06985 h...
1	[-0.015508202835917473, 0.017525529488921165, ...	[e9748376-1181-46a7-a376-d03b3baa82a5]	[{'file_path': '/content/data/2202.11983.txt',...	[Computer Science Computer Vision and Pattern...

Figure 4.2: Output of the text dataset structure and content

We will now initialize the LLM query engine.

Initializing the LLM query engine

As in *Chapter 3, Building Indexed-Based RAG with LlamaIndex, Deep Lake, and OpenAI*, we will initialize a vector store index from the collection of drone documents (documents_llm) of the dataset (ds). The GPTVectorStoreIndex.from_documents() method creates an index that increases the retrieval speed of documents based on vector similarity:

```
from llama_index.core import VectorStoreIndex
vector_store_index_llm = VectorStoreIndex.from_documents(documents_llm)
```

The as_query_engine() method configures this index as a query engine with the specific parameters, as in *Chapter 3*, for similarity and retrieval depth, allowing the system to answer queries by finding the most relevant documents:

```
vector_query_engine_llm = vector_store_index_llm.as_query_engine(similarity_
top_k=2, temperature=0.1, num_output=1024)
```

Now, the program introduces the user input.

User input for multimodal modular RAG

The goal of defining the user input in the context of the modular RAG system is to formulate a query that will effectively utilize both the text-based and image-based capabilities. This allows the system to generate a comprehensive and accurate response by leveraging multiple information sources:

```
user_input="How do drones identify a truck?"
```

In this context, the user input is the *baseline*, the starting point, or a standard query used to assess the system's capabilities. It will establish the initial frame of reference for how well the system can handle and respond to queries utilizing its available resources (e.g., text and image data from various datasets). In this example, the baseline is empirical and will serve to evaluate the system from that reference point.

Querying the textual dataset

We will run the vector query engine request as we did in *Chapter 3*:

```
import time
import textwrap
#start the timer
start_time = time.time()
llm_response = vector_query_engine_llm.query(user_input)
# Stop the timer
end_time = time.time()
# Calculate and print the execution time
elapsed_time = end_time - start_time
print(f"Query execution time: {elapsed_time:.4f} seconds")
print(textwrap.fill(str(llm_response), 100))
```

The execution time is satisfactory:

```
Query execution time: 1.5489 seconds
```

The output content is also satisfactory:

```
Drones can identify a truck using visual detection and tracking methods, which
may involve deep neural networks for performance benchmarking.
```

The program now loads the multimodal drone dataset.

Loading and visualizing the multimodal dataset

We will use the existing pubic VisDrone dataset available on Deep Lake: https://datasets.activeloop.
ai/docs/ml/datasets/visdrone-dataset/. We will *not* create a vector store but simply load the
existing dataset in memory:

```
import deeplake


dataset_path = 'hub://activeloop/visdrone-det-train'
ds = deeplake.load(dataset_path) # Returns a Deep Lake Dataset but does not
download data locally
```

The output will display a link to the online dataset that you can explore with SQL, or natural language
processing commands if you prefer, with the tools provided by Deep Lake:

```
Opening dataset in read-only mode as you don't have write permissions.
This dataset can be visualized in Jupyter Notebook by ds.visualize() or at
https://app.activeloop.ai/activeloop/visdrone-det-train

hub://activeloop/visdrone-det-train loaded successfully.
```

Let's display the summary to explore the dataset in code:

```
ds.summary()
```

The output provides useful information on the structure of the dataset:

```
Dataset(path='hub://activeloop/visdrone-det-train', read_only=True,
tensors=['boxes', 'images', 'labels'])

 tensor      htype            shape            dtype     compression
 ------      -----            -----            -----     -----------
 boxes       bbox         (6471, 1:914, 4)     float32         None
 images      image        (6471, 360:1500,
                            480:2000, 3)        uint8           jpeg
 labels      class_label  (6471, 1:914)        uint32          None
```

The structure contains images, boxes for the boundary boxes of the objects in the image, and labels describing the images and boundary boxes. Let's visualize the dataset in code:

```
ds.visualize()
```

The output shows the images and their boundary boxes:

Figure 4.3: Output showing boundary boxes

🔍 **Quick tip:** Need to see a high-resolution version of this image? Open this book in the next-gen Packt Reader or view it in the PDF/ePub copy.

🔒 **The next-gen Packt Reader** and a **free PDF/ePub copy** of this book are included with your purchase. Unlock them by scanning the QR code below or visiting https://www.packtpub.com/unlock/9781836200918.

Now, let's go further and display the content of the dataset in a pandas DataFrame to see what the images look like:

```python
import pandas as pd

# Create an empty DataFrame with the defined structure
df = pd.DataFrame(columns=['image', 'boxes', 'labels'])

# Iterate through the samples using enumerate
for i, sample in enumerate(ds):

    # Image data (choose either path or compressed representation)
    # df.loc[i, 'image'] = sample.images.path  # Store image path
    df.loc[i, 'image'] = sample.images.tobytes()  # Store compressed image data

    # Bounding box data (as a list of lists)
    boxes_list = sample.boxes.numpy(aslist=True)
    df.loc[i, 'boxes'] = [box.tolist() for box in boxes_list]

    # Label data (as a list)
    label_data = sample.labels.data()
    df.loc[i, 'labels'] = label_data['text']

df
```

The output in *Figure 4.4* shows the content of the dataset:

	image	boxes	labels
0	b'\xff\xd8\xff\xe0\x00\x10JFIF\x00\x01\x01\x00...	[[1221.0, 84.0, 16.0, 33.0], [1235.0, 71.0, 18...	[pedestrian, pedestrian, tricycle, pedestrian,...
1	b'\xff\xd8\xff\xe0\x00\x10JFIF\x00\x01\x01\x00...	[[351.0, 936.0, 305.0, 114.0], [0.0, 818.0, 22...	[car, car, car, car, car, car, car, car, car, ...
2	b'\xff\xd8\xff\xe0\x00\x10JFIF\x00\x01\x01\x00...	[[699.0, 716.0, 26.0, 54.0], [600.0, 604.0, 22...	[truck, car, van, car, car, car, car, car, car...
3	b'\xff\xd8\xff\xe0\x00\x10JFIF\x00\x01\x01\x00...	[[417.0, 77.0, 57.0, 54.0], [387.0, 109.0, 31...	[ignored regions, ignored regions, car, car, c...
4	b'\xff\xd8\xff\xe0\x00\x10JFIF\x00\x01\x01\x00...	[[794.0, 617.0, 97.0, 115.0], [803.0, 539.0, 7...	[car, car, car, car, car, car, car, car, car, ...
...	...	...	...
6466	b'\xff\xd8\xff\xe0\x00\x10JFIF\x00\x01\x01\x00...	[[683.0, 710.0, 7.0, 6.0], [681.0, 717.0, 5.0,...	[car, car, car, car, awning-tricycle, car, car...
6467	b'\xff\xd8\xff\xe0\x00\x10JFIF\x00\x01\x01\x00...	[[1761.0, 676.0, 207.0, 155.0], [1539.0, 648.0...	[car, car, car, car, car, van, car, car, car, ...
6468	b'\xff\xd8\xff\xe0\x00\x10JFIF\x00\x01\x01\x00...	[[0.0, 501.0, 105.0, 93.0], [412.0, 594.0, 172...	[car, van, pedestrian, pedestrian, pedestrian,...
6469	b'\xff\xd8\xff\xe0\x00\x10JFIF\x00\x01\x01\x00...	[[200.0, 604.0, 112.0, 64.0], [311.0, 560.0, 1...	[van, truck, van, car, car, van, others, car, ...
6470	b'\xff\xd8\xff\xe0\x00\x10JFIF\x00\x01\x01\x00...	[[60.0, 144.0, 16.0, 21.0], [51.0, 120.0, 15.0...	[people, pedestrian, others, people, people, m...

6471 rows × 3 columns

Figure 4.4: Excerpt of the VisDrone dataset

There are 6,471 rows of images in the dataset and 3 columns:

- The `image` column contains the image. The format of the image in the dataset, as indicated by the byte sequence b'\xff\xd8\xff\xe0\x00\x10JFIF\x00\x01\x01\x00...', is JPEG. The bytes b'\xff\xd8\xff\xe0' specifically signify the start of a JPEG image file.
- The `boxes` column contains the coordinates and dimensions of bounding boxes in the image, which are normally in the format [x, y, width, height].
- The `labels` column contains the label of each bounding box in the `boxes` column.

We can display the list of labels for the images:

```
labels_list = ds.labels.info['class_names']
labels_list
```

The output provides the list of labels, which defines the scope of the dataset:

```
['ignored regions',
 'pedestrian',
 'people',
 'bicycle',
 'car',
 'van',
 'truck',
 'tricycle',
 'awning-tricycle',
 'bus',
 'motor',
 'others']
```

With that, we have successfully loaded the dataset and will now explore the multimodal dataset structure.

Navigating the multimodal dataset structure

In this section, we will select an image and display it using the dataset's image column. To this image, we will then add the bounding boxes of a label that we will choose. The program first selects an image.

Selecting and displaying an image

We will select the first image in the dataset:

```
# choose an image
ind=0
image = ds.images[ind].numpy() # Fetch the first image and return a numpy array
```

Now, let's display it with no bounding boxes:

```
import deeplake
from IPython.display import display
from PIL import Image
import cv2  # Import OpenCV

image = ds.images[0].numpy()

# Convert from BGR to RGB (if necessary)
image_rgb = cv2.cvtColor(image, cv2.COLOR_BGR2RGB)

# Create PIL Image and display
img = Image.fromarray(image_rgb)
display(img)
```

The image displayed contains trucks, pedestrians, and other types of objects:

Figure 4.5: Output displaying objects

Now that the image is displayed, the program will add bounding boxes.

Adding bounding boxes and saving the image

We have displayed the first image. The program will then fetch all the labels for the selected image:

```
labels = ds.labels[ind].data() # Fetch the labels in the selected image
print(labels)
```

The output displays value, which contains the numerical indices of a label, and text, which contains the corresponding text labels of a label:

```
{'value': array([1, 1, 7, 1, 1, 1, 1, 6, 6, 6, 6, 6, 6, 6, 6, 6, 6, 1, 6, 6, 6,
6,
      1, 1, 1, 1, 1, 1, 6, 6, 3, 6, 6, 1, 1, 1, 1, 1, 1, 1, 1, 1, 1, 1,
```

```
      1, 6, 6, 6], dtype=uint32), 'text': ['pedestrian', 'pedestrian',
 'tricycle', 'pedestrian', 'pedestrian', 'pedestrian', 'pedestrian', 'truck',
 'truck', 'truck', 'truck', 'truck', 'truck', 'truck', 'truck', 'truck',
 'truck', 'pedestrian', 'truck', 'truck', 'truck', 'truck', 'pedestrian',
 'pedestrian', 'pedestrian', 'pedestrian', 'pedestrian', 'pedestrian', 'truck',
 'truck', 'bicycle', 'truck', 'truck', 'pedestrian', 'pedestrian', 'pedestrian',
 'pedestrian', 'pedestrian', 'pedestrian', 'pedestrian', 'pedestrian',
 'pedestrian', 'pedestrian', 'pedestrian', 'pedestrian', 'truck', 'truck',
 'truck']}
```

We can display the values and the corresponding text in two columns:

```
values = labels['value']
text_labels = labels['text']

# Determine the maximum text label length for formatting
max_text_length = max(len(label) for label in text_labels)

# Print the header
print(f"{'Index':<10}{'Label':<{max_text_length + 2}}")
print("-" * (10 + max_text_length + 2))  # Add a separator line

# Print the indices and labels in two columns
for index, label in zip(values, text_labels):
    print(f"{index:<10}{label:<{max_text_length + 2}}")
```

The output gives us a clear representation of the content of the labels of an image:

```
Index     Label
---------------------
1         pedestrian
1         pedestrian
7         tricycle
1         pedestrian
1         pedestrian
1         pedestrian
1         pedestrian
6         truck
6         truck    …
```

We can group the class names (labels in plain text) of the images:

```
ds.labels[ind].info['class_names'] # class names of the selected image
```

We can now group and display all the labels that describe the image:

```
ds.labels[ind].info['class_names'] #class names of the selected image
```

We can see all the classes the image contains:

```
['ignored regions',
 'pedestrian',
 'people',
 'bicycle',
 'car',
 'van',
 'truck',
 'tricycle',
 'awning-tricycle',
 'bus',
 'motor',
 'others']
```

The number of label classes sometimes exceeds what a human eye can see in an image.

Let's now add bounding boxes. We first create a function to add the bounding boxes, display them, and save the image:

```
def display_image_with_bboxes(image_data, bboxes, labels, label_name, ind=0):
    #Displays an image with bounding boxes for a specific label.

    image_bytes = io.BytesIO(image_data)
    img = Image.open(image_bytes)

    # Extract class names specifically for the selected image
    class_names = ds.labels[ind].info['class_names']

    # Filter for the specific label (or display all if class names are missing)
    if class_names is not None:
        try:
            label_index = class_names.index(label_name)
            relevant_indices = np.where(labels == label_index)[0]
        except ValueError:
            print(f"Warning: Label '{label_name}' not found. Displaying all
boxes.")
            relevant_indices = range(len(labels))
    else:
        relevant_indices = []   # No labels found, so display no boxes
```

```
# Draw bounding boxes
draw = ImageDraw.Draw(img)
for idx, box in enumerate(bboxes):  # Enumerate over bboxes
    if idx in relevant_indices:   # Check if this box is relevant
        x1, y1, w, h = box
        x2, y2 = x1 + w, y1 + h
        draw.rectangle([x1, y1, x2, y2], outline="red", width=2)
        draw.text((x1, y1), label_name, fill="red")
# Save the image
save_path="boxed_image.jpg"
img.save(save_path)
display(img)
```

We can add the bounding boxes for a specific label. In this case, we selected the "truck" label:

```
import io
from PIL import ImageDraw
# Fetch labels and image data for the selected image
labels = ds.labels[ind].data()['value']
image_data = ds.images[ind].tobytes()
bboxes = ds.boxes[ind].numpy()
ibox="truck" # class in image
# Display the image with bounding boxes for the label chosen
display_image_with_bboxes(image_data, bboxes, labels, label_name=ibox)
```

The image displayed now contains the bounding boxes for trucks:

Figure 4.6: Output displaying bounding boxes

🔍 **Quick tip:** Need to see a high-resolution version of this image? Open this book in the next-gen Packt Reader or view it in the PDF/ePub copy.

🔒 **The next-gen Packt Reader** and a **free PDF/ePub copy** of this book are included with your purchase. Unlock them by scanning the QR code below or visiting `https://www.packtpub.com/unlock/9781836200918`.

Let's now activate a query engine to retrieve and obtain a response.

Building a multimodal query engine

In this section, we will query the VisDrone dataset and retrieve an image that fits the user input we entered in the *User input for multimodal modular RAG* section of this notebook. To achieve this goal, we will:

1. Create a vector index for each row of the df DataFrame containing the images, boxing data, and labels of the VisDrone dataset.
2. Create a query engine that will query the text data of the dataset, retrieve relevant image information, and provide a text response.
3. Parse the nodes of the response to find the keywords related to the user input.
4. Parse the nodes of the response to find the source image.
5. Add the bounding boxes of the source image to the image.
6. Save the image.

Creating a vector index and query engine

The code first creates a document that will be processed to create a vector store index for the multimodal drone dataset. The df DataFrame we created in the *Loading and visualizing the multimodal dataset* section of the notebook on GitHub does not have unique indices or embeddings. We will create them in memory with LlamaIndex.

The program first assigns a unique ID to the DataFrame:

```
# The DataFrame is named 'df'
df['doc_id'] = df.index.astype(str)  # Create unique IDs from the row indices
```

This line adds a new column to the `df` DataFrame called `doc_id`. It assigns unique identifiers to each row by converting the DataFrame's row indices to strings. An empty list named `documents` is initialized, which we will use to create a vector index:

```
# Create documents (extract relevant text for each image's labels)
documents = []
```

Now, the `iterrows()` method iterates through each row of the DataFrame, generating a sequence of index and row pairs:

```
for _, row in df.iterrows():
    text_labels = row['labels'] # Each label is now a string
    text = " ".join(text_labels) # Join text labels into a single string
    document = Document(text=text, doc_id=row['doc_id'])
    documents.append(document)
```

`documents` is appended with all the records in the dataset, and a DataFrame is created:

```
# The DataFrame is named 'df'
df['doc_id'] = df.index.astype(str)  # Create unique IDs from the row indices

# Create documents (extract relevant text for each image's labels)
documents = []
for _, row in df.iterrows():
    text_labels = row['labels'] # Each label is now a string
    text = " ".join(text_labels) # Join text labels into a single string
    document = Document(text=text, doc_id=row['doc_id'])
    documents.append(document)
```

The documents are now ready to be indexed with `GPTVectorStoreIndex`:

```
from llama_index.core import GPTVectorStoreIndex
vector_store_index = GPTVectorStoreIndex.from_documents(documents)
```

The dataset is then seamlessly equipped with indices that we can visualize in the index dictionary:

```
vector_store_index.index_struct
```

The output shows that an index has now been added to the dataset:

```
IndexDict(index_id='4ec313b4-9a1a-41df-a3d8-a4fe5ff6022c', summary=None,
nodes_dict={'5e547c1d-0d65-4de6-b33e-a101665751e6': '5e547c1d-0d65-4de6-b33e-
a101665751e6', '05f73182-37ed-4567-a855-4ff9e8ae5b8c': '05f73182-37ed-4567-
a855-4ff9e8ae5b8c'
```

We can now run a query on the multimodal dataset.

Running a query on the VisDrone multimodal dataset

We now set vector_store_index as the query engine, as we did in the *Vector store index query engine* section in *Chapter 3*:

```
vector_query_engine = vector_store_index.as_query_engine(similarity_top_k=1,
temperature=0.1, num_output=1024)
```

We can also run a query on the dataset of drone images, just as we did in *Chapter 3* on an LLM dataset:

```
import time
start_time = time.time()
response = vector_query_engine.query(user_input)
# Stop the timer
end_time = time.time()
# Calculate and print the execution time
elapsed_time = end_time - start_time
print(f"Query execution time: {elapsed_time:.4f} seconds")
```

The execution time is satisfactory:

```
Query execution time: 1.8461 seconds
```

We will now examine the text response:

```
print(textwrap.fill(str(response), 100))
```

We can see that the output is logical and therefore satisfactory.

Drones use various sensors such as cameras, LiDAR, and GPS to identify and track objects like trucks.

Processing the response

We will now parse the nodes in the response to find the unique words in the response and select one for this notebook:

```
from itertools import groupby

def get_unique_words(text):
    text = text.lower().strip()
    words = text.split()
    unique_words = [word for word, _ in groupby(sorted(words))]
    return unique_words

for node in response.source_nodes:
    print(node.node_id)
    # Get unique words from the node text:
    node_text = node.get_text()
```

```
unique_words = get_unique_words(node_text)
print("Unique Words in Node Text:", unique_words)
```

We found a unique word ('truck') and its unique index, which will lead us directly to the image of the source of the node that generated the response:

```
1af106df-c5a6-4f48-ac17-f953dffd2402
Unique Words in Node Text: ['truck']
```

We could select more words and design this function in many different ways depending on the specifications of each project.

We will now search for the image by going through the source nodes, just as we did for an LLM dataset in the *Query response and source* section of the previous chapter. Multimodal vector stores and querying frameworks are flexible. Once we learn how to perform retrievals on an LLM and a multimodal dataset, we are ready for anything that comes up!

Let's select and process the information related to an image.

Selecting and processing the image of the source node

Before running the image retrieval and displaying function, let's first delete the image we displayed in the *Adding bounding boxes and saving the image* section of this notebook to make sure we are working on a new image:

```
# deleting any image previously saved
!rm /content/boxed_image.jpg
```

We are now ready to search for the source image, call the bounding box, and display and save the function we defined earlier:

```
display_image_with_bboxes(image_data, bboxes, labels, label_name=ibox)
```

The program now goes through the source nodes with the keyword "truck" search, applies the bounding boxes, and displays and saves the image:

```
import io
from PIL import Image


def process_and_display(response, df, ds, unique_words):
    """Processes nodes, finds corresponding images in dataset, and displays
them with bounding boxes.

    Args:
        response: The response object containing source nodes.
        df: The DataFrame with doc_id information.
        ds: The dataset containing images, labels, and boxes.
        unique_words: The list of unique words for filtering.
```

```
            """

    ...

            if i == row_index:
                image_bytes = io.BytesIO(sample.images.tobytes())
                img = Image.open(image_bytes)

                labels = ds.labels[i].data()['value']
                image_data = ds.images[i].tobytes()
                bboxes = ds.boxes[i].numpy()
                ibox = unique_words[0]   # class in image

                display_image_with_bboxes(image_data, bboxes, labels, label_
    name=ibox)

    # Assuming you have your 'response', 'df', 'ds', and 'unique_words' objects
    prepared:
    process_and_display(response, df, ds, unique_words)
```

The output is satisfactory:

Figure 4.7: Displayed satisfactory output

Multimodal modular summary

We have built a multimodal modular program step by step that we can now assemble in a summary. We will create a function to display the source image of the response to the user input, then print the user input and the LLM output, and display the image.

First, we create a function to display the source image saved by the multimodal retrieval engine:

```
# 1.user input=user_input
print(user_input)
# 2.LLM response
print(textwrap.fill(str(llm_response), 100))
# 3.Multimodal response
image_path = "/content/boxed_image.jpg"
display_source_image(image_path)
```

Then, we can display the user input, the LLM response, and the multimodal response. The output first displays the textual responses (user input and LLM response):

```
How do drones identify a truck?
Drones can identify a truck using visual detection and tracking methods, which
may involve deep neural networks for performance benchmarking.
```

Then, the image is displayed with the bounding boxes for trucks in this case:

Figure 4.8: Output displaying boundary boxes

🔍 **Quick tip:** Need to see a high-resolution version of this image? Open this book in the next-gen Packt Reader or view it in the PDF/ePub copy.

🔒 **The next-gen Packt Reader** and a **free PDF/ePub copy** of this book are included with your purchase. Unlock them by scanning the QR code below or visiting https://www. packtpub.com/unlock/9781836200918.

By adding an image to a classical LLM response, we augmented the output. Multimodal RAG output augmentation will enrich generative AI by adding information to both the input and output. However, as for all AI programs, designing a performance metric requires efficient image recognition functionality.

Performance metric

Measuring the performance of a multimodal modular RAG requires two types of measurements: text and image. Measuring text is straightforward. However, measuring images is quite a challenge. Analyzing the image of a multimodal response is quite different. We extracted a keyword from the multimodal query engine. We then parsed the response for a source image to display. However, we will need to build an innovative approach to evaluate the source image of the response. Let's begin with the LLM performance.

LLM performance metric

LlamaIndex seamlessly called an OpenAI model through its query engine, such as GPT-4, for example, and provided text content in its response. For text responses, we will use the same cosine similarity metric as in the *Evaluating the output with cosine similarity* section in *Chapter 2*, and the *Vector store index query engine* section in *Chapter 3*.

The evaluation function uses `sklearn` and `sentence_transformers` to evaluate the similarity between two texts—in this case, an input and an output:

```
from sklearn.feature_extraction.text import TfidfVectorizer
from sklearn.metrics.pairwise import cosine_similarity

from sentence_transformers import SentenceTransformer
model = SentenceTransformer('all-MiniLM-L6-v2')

def calculate_cosine_similarity_with_embeddings(text1, text2):
    embeddings1 = model.encode(text1)
    embeddings2 = model.encode(text2)
    similarity = cosine_similarity([embeddings1], [embeddings2])
    return similarity[0][0]
```

We can now calculate the similarity between our baseline user input and the initial LLM response obtained:

```
llm_similarity_score = calculate_cosine_similarity_with_embeddings(user_input,
str(llm_response))
print(user_input)
print(llm_response)
print(f"Cosine Similarity Score: {llm_similarity_score:.3f}")
```

The output displays the user input, the text response, and the cosine similarity between the two texts:

```
How do drones identify a truck?
How do drones identify a truck?
```

```
Drones can identify a truck using visual detection and tracking methods, which
may involve deep neural networks for performance benchmarking.
Cosine Similarity Score: 0.691
```

The output is satisfactory. But we now need to design a way to measure the multimodal performance.

Multimodal performance metric

To evaluate the image returned, we cannot simply rely on the labels in the dataset. For small datasets, we can manually check the image, but when a system scales, automation is required. In this section, we will use the computer vision features of GPT-4o to analyze an image, parse it to find the objects we are looking for, and provide a description of that image. Then, we will apply cosine similarity to the description provided by GPT-4o and the label it is supposed to contain. GPT-4o is a multimodal generative AI model.

Let's first encode the image to simplify data transmission to GPT-4o. Base64 encoding converts binary data (like images) into ASCII characters, which are standard text characters. This transformation is crucial because it ensures that the image data can be transmitted over protocols (like HTTP) that are designed to handle text data smoothly. It also avoids issues related to binary data transmission, such as data corruption or interpretation errors.

The program encodes the source image using Python's `base64` module:

```
import base64

IMAGE_PATH = "/content/boxed_image.jpg"

# Open the image file and encode it as a base64 string
def encode_image(image_path):
    with open(image_path, "rb") as image_file:
        return base64.b64encode(image_file.read()).decode("utf-8")

base64_image = encode_image(IMAGE_PATH)
```

We now create an OpenAI client and set the model to `gpt-4o`:

```
from openai import OpenAI
#Set the API key for the client
client = OpenAI(api_key=openai.api_key)
MODEL="gpt-4o"
```

The unique word will be the result of the LLM query to the multimodal dataset we obtained by parsing the response:

```
u_word=unique_words[0]
print(u_word)
```

We can now submit the image to OpenAI GPT-4o:

```python
response = client.chat.completions.create(
    model=MODEL,
    messages=[
        {"role": "system", "content": f"You are a helpful assistant that
analyzes images that contain {u_word}."},
        {"role": "user", "content": [
            {"type": "text", "text": f"Analyze the following image, tell me if
there is one {u_word} or more in the bounding boxes and analyze them:"},
            {"type": "image_url", "image_url": {
                "url": f"data:image/png;base64,{base64_image}"}
            }
        ]}
    ],
    temperature=0.0,
)
response_image = response.choices[0].message.content
print(response_image)
```

We instructed the system and user roles to analyze images looking for our target label, u_word—in this case, truck. We then submitted the source node image to the model. The output that describes the image is satisfactory:

```
The image contains two trucks within the bounding boxes. Here is the analysis
of each truck:


1. **First Truck (Top Bounding Box)**:
   - The truck appears to be a flatbed truck.
   - It is loaded with various materials, possibly construction or industrial
supplies.
   - The truck is parked in an area with other construction materials and
equipment.


2. **Second Truck (Bottom Bounding Box)**:
   - This truck also appears to be a flatbed truck.
   - It is carrying different types of materials, similar to the first truck.
   - The truck is situated in a similar environment, surrounded by construction
materials and equipment.


Both trucks are in a construction or industrial area, likely used for
transporting materials and equipment.
```

We can now submit this response to the cosine similarity function by first adding an "s" to align with multiple trucks in a response:

```
resp=u_word+"s"
multimodal_similarity_score = calculate_cosine_similarity_with_embeddings(resp,
str(response_image))
print(f"Cosine Similarity Score: {multimodal_similarity_score:.3f}")
```

The output describes the image well but contains many other descriptions beyond the word "truck," which limits its similarity to the input requested:

```
Cosine Similarity Score: 0.505
```

A human observer might approve the image and the LLM response. However, even if the score was very high, the issue would be the same. Complex images are challenging to analyze in detail and with precision, although progress is continually made. Let's now calculate the overall performance of the system.

Multimodal modular RAG performance metric

To obtain the overall performance of the system, we will divide the sum of the LLM response and the two multimodal response performances by 2:

```
score=(llm_similarity_score+multimodal_similarity_score)/2
print(f"Multimodal, Modular Score: {score:.3f}")
```

The result shows that although a human who observes the results may be satisfied, it remains difficult to automatically assess the relevance of a complex image:

```
Multimodal, Modular Score: 0.598
```

The metric can be improved because a human observer sees that the image is relevant. This explains why the top AI agents, such as ChatGPT, Gemini, and Bing Copilot, always have a feedback process that includes thumbs up and thumbs down.

Let's now sum up the chapter and gear up to explore how RAG can be improved even further with human feedback.

Summary

This chapter introduced us to the world of multimodal modular RAG, which uses distinct modules for different data types (text and image) and tasks. We leveraged the functionality of LlamaIndex, Deep Lake, and OpenAI, which we explored in the previous chapters. The Deep Lake VisDrone dataset further introduced us to drone technology for analyzing images and identifying objects. The dataset contained images, labels, and bounding box information. Working on drone technology involves multimodal data, encouraging us to develop skills that we can use across many domains, such as wildlife tracking, streamlining commercial deliveries, and making safer infrastructure inspections.

We built a multimodal modular RAG-driven generative AI system. The first step was to define a baseline user query for both LLM and multimodal queries. We began by querying the Deep Lake textual dataset that we implemented in *Chapter 3*. LlamaIndex seamlessly ran a query engine to retrieve, augment, and generate a response. Then, we loaded the Deep Lake VisDrone dataset and indexed it in memory with LlamaIndex to create an indexed vector search retrieval pipeline. We queried it through LlamaIndex, which used an OpenAI model such as GPT-4 and parsed the text generated for a keyword. Finally, we searched the source nodes of the response to find the source image, display it, and merge the LLM and image responses into an augmented output. We applied cosine similarity to the text response. Evaluating the image was challenging, so we first ran image recognition with GPT-4o on the image retrieved to obtain a text to which we applied cosine similarity.

The journey into multimodal modular RAG-driven generative AI took us deep into the cutting edge of AI. Building a complex system was good preparation for real-life AI projects, which often require implementing multisource, multimodal, and unstructured data, leading to modular, complex systems. Thanks to transparent access to the source of a response, the complexity of RAG can be harnessed, controlled, and improved. We will see how we can leverage the transparency of the sources of a response to introduce human feedback to improve AI. The next chapter will take us further into transparency and precision in AI.

Questions

Answer the following questions with *Yes* or *No*:

1. Does multimodal modular RAG handle different types of data, such as text and images?
2. Are drones used solely for agricultural monitoring and aerial photography?
3. Is the Deep Lake VisDrone dataset used in this chapter for textual data only?
4. Can bounding boxes be added to drone images to identify objects such as trucks and pedestrians?
5. Does the modular system retrieve both text and image data for query responses?
6. Is building a vector index necessary for querying the multimodal VisDrone dataset?
7. Are the retrieved images processed without adding any labels or bounding boxes?
8. Is the multimodal modular RAG performance metric based only on textual responses?
9. Can a multimodal system such as the one described in this chapter handle only drone-related data?
10. Is evaluating images as easy as evaluating text in multimodal RAG?

References

- LlamaIndex: https://docs.llamaindex.ai/en/stable/
- Activeloop Deep Lake: https://docs.activeloop.ai/
- OpenAI: https://platform.openai.com/docs/overview

Further reading

- Retrieval-Augmented Multimodal Language Modeling, Yasunaga et al. (2023), `https://arxiv.org/pdf/2211.12561`

Unlock this book's exclusive benefits now

This book comes with additional benefits designed to elevate your learning experience.

Note: Have your purchase invoice ready before you begin. `https://www.packtpub.com/unlock/9781836200918`

5

Boosting RAG Performance with Expert Human Feedback

Human feedback (HF) is not just useful for generative AI—it's essential, especially when it comes to models using RAG. A generative AI model uses information from datasets with various documents during training. The data that trained the AI model is set in stone in the model's parameters; we can't change it unless we train it again. However, in the world of retrieval-based text and multimodal datasets, there is information we can see and tweak. That is where HF comes in. By providing feedback on what the AI model pulls from its datasets, HF can directly influence the quality of its future responses. Engaging with this process makes humans an active player in the RAG's development. It adds a new dimension to AI projects: adaptive RAG.

We have explored and implemented naïve, advanced, and modular RAG so far. Now, we will add adaptive RAG to our generative AI toolbox. We know that even the best generative AI system with the best metrics cannot convince a dissatisfied user that it is helpful if it isn't. We will introduce adaptive RAG with an HF loop. The system thus becomes adaptive because the documents used for retrieval are updated. Integrating HF in RAG leads to a pragmatic hybrid approach because it involves humans in an otherwise automated generative process. We will thus leverage HF, which we will use to build a hybrid adaptive RAG program in Python from scratch, going through the key steps of building a RAG-driven generative AI system from the ground up. By the end of this chapter, you will have a theoretical understanding of the adaptive RAG framework and practical experience in building an AI model based on HF.

This chapter covers the following topics:

- Defining the adaptive RAG ecosystem
- Applying adaptive RAG to augmented retrieval queries
- Automating augmented generative AI inputs with HF
- Automating end-user feedback rankings to trigger expert HF
- Creating an automated feedback system for a human expert
- Integrating HF with adaptive RAG for GPT-4o

Let's begin by defining adaptive RAG.

Adaptive RAG

No, RAG cannot solve all our problems and challenges. RAG, just like any generative model, can also produce irrelevant and incorrect output! RAG might be a useful option, however, because we feed pertinent documents to the generative AI model that inform its responses. Nonetheless, the quality of RAG outputs depends on the accuracy and relevance of the underlying data, which calls for verification! That's where adaptive RAG comes in. Adaptive RAG introduces human, real-life, pragmatic feedback that will improve a RAG-driven generative AI ecosystem.

The core information in a generative AI model is parametric (stored as weights). But in the context of RAG, this data can be visualized and controlled, as we saw in *Chapter 2, RAG Embedding Vector Stores with Deep Lake and OpenAI*. Despite this, challenges remain; for example, the end-user might write fuzzy queries, or the RAG data retrieval might be faulty. An HF process is, therefore, highly recommended to ensure the system's reliability.

Figure 1.3 from *Chapter 1, Why Retrieval Augmented Generation?*, represents the complete RAG framework and ecosystem. Let's zoom in on the adaptive RAG ecosystem and focus on the key processes that come into play, as shown in the following figure:

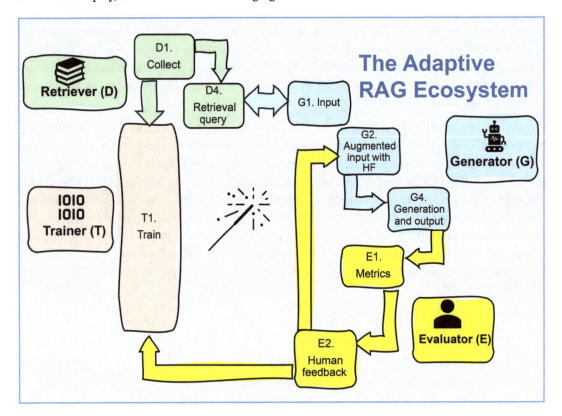

Figure 5.1: A variant of an adaptive RAG ecosystem

The variant of an adaptive RAG ecosystem in this chapter includes the following components, as shown in *Figure 5.1*, for the retriever:

- **D1: Collect and process** Wikipedia articles on LLMs by fetching and cleaning the data
- **D4: Retrieval query** to query the retrieval dataset

The generator's components are:

- **G1: Input** entered by an end-user
- **G2: Augmented input with HF** that will augment the user's initial input and **prompt engineering** to configure the GPT-4o model's prompt
- **G4: Generation and output** to run the generative AI model and obtain a response

The evaluator's components are:

- **E1: Metrics** to apply a cosine similarity measurement
- **E2: Human feedback** to obtain and process the ultimate measurement of a system through end-user and expert feedback

In this chapter, we will illustrate adaptive RAG by building a hybrid adaptive RAG program in Python on Google Colab. We will build this program from scratch to acquire a clear understanding of an adaptive process, which may vary depending on a project's goals, but the underlying principles remain the same. Through this hands-on experience, you will learn how to develop and customize a RAG system when a ready-to-use one fails to meet the users' expectations. This is important because human users can be dissatisfied with a response no matter what the performance metrics show. We will also explore the incorporation of human user rankings to gather expert feedback on our RAG-driven generative AI system. Finally, we will implement an automated ranking system that will decide how to augment the user input for the generative model, offering practical insights into how a RAG-driven system can be successfully implemented in a company.

We will develop a proof of concept for a hypothetical company called *Company C*. This company would like to deploy a conversational agent that explains what AI is. The goal is for the employees of this company to understand the basic terms, concepts, and applications of AI. The ML engineer in charge of this RAG-driven generative AI example would like future users to acquire a better knowledge of AI while implementing other AI projects across the sales, production, and delivery domains.

Company C currently faces serious issues with customer support. With a growing number of products and services, their product line of smartphones of the C-phone series has been experiencing technical problems with too many customer requests. The IT department would like to set up a conversational agent for these customers. However, the teams are not convinced. The IT department has thus decided to first set up a conversational agent to explain what an LLM is and how it can be helpful in the C-phone series customer support service.

The program will be hybrid and adaptive to fulfill the needs of Company C:

- **Hybrid:** Real-life scenarios go beyond theoretical frameworks and configurations. The system is hybrid because we are integrating HF within the retrieval process that can be processed in real time. However, we will not parse the content of the documents with a keyword alone. We will label the documents (which are Wikipedia URLs in this case), which can be done automatically, controlled, and improved *by a human*, if necessary. As we show in this chapter, some documents will be replaced by human-expert feedback and relabeled. The program will automatically retrieve human-expert feedback documents and raw retrieved documents to form a hybrid (human-machine) *dynamic* RAG system.
- **Adaptive:** We will introduce human user ranking, expert feedback, and automated document re-ranking. This HF loop takes us deep into modular RAG and adaptive RAG. Adaptive RAG leverages the flexibility of a RAG system to adapt its responses to the queries. In this case, we want HF to be triggered to improve the quality of the output.

Real-life projects will inevitably require an ML engineer to go beyond the boundaries of pre-determined categories. Pragmatism and necessity encourage creative and innovative solutions. For example, for the hybrid, dynamic, and adaptive aspects of the system, ML engineers could imagine any process that works with any type of algorithm: classical software functions, ML clustering algorithms, or any function that works. In real-life AI, what works, works!

It's time to build a proof of concept to show Company C's management how hybrid adaptive RAG-driven generative AI can successfully help their teams by:

- Proving that AI can work with a proof of concept before scaling and investing in a project
- Showing that an AI system can be customized for a specific project
- Developing solid ground-up skills to face any AI challenge
- Building the company's data governance and control of AI systems
- Laying solid grounds to scale the system by solving the problems that will come up during the proof of concept

Let's go to our keyboards!

Building hybrid adaptive RAG in Python

Let's now start building the proof of concept of a hybrid adaptive RAG-driven generative AI configuration. Open `Adaptive_RAG.ipynb` on GitHub. We will focus on HF and, as such, will not use an existing framework. We will build our own pipeline and introduce HF.

As established earlier, the program is divided into three separate parts: the **retriever**, **generator**, and **evaluator** functions, which can be separate agents in a real-life project's pipeline. Try to separate these functions from the start because, in a project, several teams might be working in parallel on separate aspects of the RAG framework.

The titles of each of the following sections correspond exactly to the names of each section in the program on GitHub. The retriever functionality comes first.

1. Retriever

We will first outline the initial steps required to set up the environment for a RAG-driven generative AI model. This process begins with the installation of essential software components and libraries that facilitate the retrieval and processing of data. We specifically cover the downloading of crucial files and the installation of packages needed for effective data retrieval and web scraping.

1.1. Installing the retriever's environment

Let's begin by downloading grequests.py from the commons directory of the GitHub repository. This repository contains resources that can be common to several programs in the repository, thus avoiding redundancy.

The download is standard and built around the request:

```
url = "https://raw.githubusercontent.com/Denis2054/RAG-Driven-Generative-AI/
main/commons/grequests.py"
output_file = "grequests.py"
```

We will only need two packages for the retriever since we are building a RAG-driven generative AI model from scratch. We will install:

- requests, the HTTP library to retrieve Wikipedia documents:

  ```
  !pip install requests==2.32.3
  ```

- beautifulsoup4, to scrape information from web pages:

  ```
  !pip install beautifulsoup4==4.12.3
  ```

We now need a dataset.

1.2.1. Preparing the dataset

For this proof of concept, we will retrieve Wikipedia documents by scraping them through their URLs. The dataset will contain automated or human-crafted labels for each document, which is the first step toward indexing the documents of a dataset:

```
import requests
from bs4 import BeautifulSoup
import re

# URLs of the Wikipedia articles mapped to keywords
```

```
urls = {
    "prompt engineering": "https://en.wikipedia.org/wiki/Prompt_engineering",
    "artificial intelligence":"https://en.wikipedia.org/wiki/Artificial_
intelligence",
    "llm": "https://en.wikipedia.org/wiki/Large_language_model",
    "llms": "https://en.wikipedia.org/wiki/Large_language_model"
}
```

One or more labels precede each URL. This approach might be sufficient for a relatively small dataset.

For specific projects, including a proof of concept, this approach can provide a solid first step to go from naïve RAG (content search with keywords) to searching a dataset with indexes (the labels in this case). We now have to process the data.

1.2.2. Processing the data

We first apply a standard scraping and text-cleaning function to the document that will be retrieved:

```
def fetch_and_clean(url):
    # Fetch the content of the URL
    response = requests.get(url)
    soup = BeautifulSoup(response.content, 'html.parser')

    # Find the main content of the article, ignoring side boxes and headers
    content = soup.find('div', {'class': 'mw-parser-output'})

    # Remove less relevant sections such as "See also", "References", etc.
    for section_title in ['References', 'Bibliography', 'External links', 'See
also']:
        section = content.find('span', {'id': section_title})
        if section:
            for sib in section.parent.find_next_siblings():
                sib.decompose()
            section.parent.decompose()

    # Focus on extracting and cleaning text from paragraph tags only
    paragraphs = content.find_all('p')
    cleaned_text = ' '.join(paragraph.get_text(separator=' ', strip=True) for
paragraph in paragraphs)
    cleaned_text = re.sub(r'\[\d+\]', '', cleaned_text)  # Remove citation
markers like [1], [2], etc.

    return cleaned_text
```

The code fetches the document's content based on its URL, which is, in turn, based on its label. This straightforward approach may satisfy a project's needs depending on its goals. An ML engineer or developer must always be careful not to overload a system with costly and unprofitable functions. Moreover, labeling website URLs can guide a retriever pipeline to the correct locations to process data, regardless of the techniques (load balancing, API call optimization, etc.) applied. In the end, each project or sub-project will require one or several techniques, depending on its specific needs.

Once the fetching and cleaning function is ready, we can implement the retrieval process for the user's input.

1.3. Retrieval process for user input

The first step here involves identifying a keyword within the user's input. The function process_query takes two parameters: user_input and num_words. The number of words to retrieve is restricted by factors like the input limitations of the model, cost considerations, and overall system performance:

```
import textwrap
def process_query(user_input, num_words):
    user_input = user_input.lower()
    # Check for any of the specified keywords in the input
    matched_keyword = next((keyword for keyword in urls if keyword in user_
input), None)
```

Upon finding a match between a keyword in the user input and the keywords associated with URLs, the following functions for fetching and cleaning the data are triggered:

```
if matched_keyword:
    print(f"Fetching data from: {urls[matched_keyword]}")
    cleaned_text = fetch_and_clean(urls[matched_keyword])

    # Limit the display to the specified number of words from the cleaned text
    words = cleaned_text.split()  # Split the text into words
    first_n_words = ' '.join(words[:num_words])  # Join the first n words into
a single string
```

The num_words parameter helps in chunking the text. While this basic approach may work for use cases with a manageable volume of data, it's recommended to embed the data into vectors for more complex scenarios.

The cleaned and truncated text is then formatted for display:

```
    # Wrap the first n words to 80 characters wide for display
    wrapped_text = textwrap.fill(first_n_words, width=80)
    print("\nFirst {} words of the cleaned text:".format(num_words))
    print(wrapped_text)  # Print the first n words as a well-formatted
paragraph
```

```
    # Use the exact same first_n_words for the GPT-4 prompt to ensure
consistency
    prompt = f"Summarize the following information about {matched_keyword}:\
n{first_n_words}"
    wrapped_prompt = textwrap.fill(prompt, width=80)  # Wrap prompt text
    print("\nPrompt for Generator:", wrapped_prompt)

    # Return the specified number of words
    return first_n_words
else:
    print("No relevant keywords found. Please enter a query related to 'LLM',
'LLMs', or 'Prompt Engineering'.")
    return None
```

Note that the function ultimately returns the first n words, providing a concise and relevant snippet of information based on the user's query. This design allows the system to manage data retrieval efficiently while also maintaining user engagement.

2. Generator

The generator ecosystem contains several components, several of which overlap with the retriever functions and user interfaces in the RAG-driven generative AI frameworks:

- **2.1. Adaptive RAG selection based on human rankings:** This will be based on the ratings of a user panel over time. In a real-life pipeline, this functionality could be a separate program.

- **2.2. Input:** In a real-life project, a **user interface (UI)** will manage the input. This interface and the associated process should be carefully designed in collaboration with the users, ideally in a workshop setting where their needs and preferences can be fully understood.

- **2.3. Mean ranking simulation scenario:** Calculating the mean value of the user evaluation scores and functionality.

- **2.4. Checking the input before running the generator:** Displaying the input.

- **2.5. Installing the generative AI environment:** The installation of the generative AI model's environment, in this case, OpenAI, can be part of another environment in the pipeline in which other team members may be working, implementing, and deploying in production independently of the retriever functionality.

- **2.6. Content generation:** In this section of the program, an OpenAI model will process the input and provide a response that will be evaluated by the evaluator.

Let's begin by describing the adaptive RAG system.

2.1. Integrating HF-RAG for augmented document inputs

The dynamic nature of information retrieval and the necessity for contextually relevant data augmentation in generative AI models require a flexible system capable of adapting to varying levels of input quality. We introduce an **adaptive RAG selection system**, which employs HF scores to determine the optimal retrieval strategy for document implementation within the RAG ecosystem. Adaptive functionality takes us beyond naïve RAG and constitutes a hybrid RAG system.

Human evaluators assign mean scores ranging from 1 to 5 to assess the relevance and quality of documents. These scores trigger distinct operational modes, as shown in the following figure:

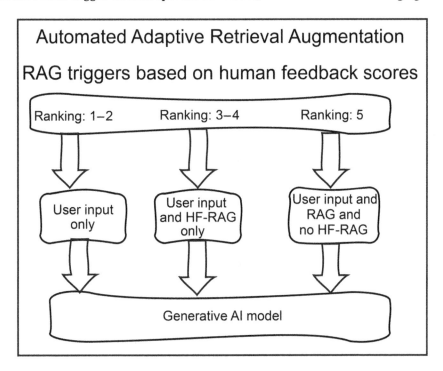

Figure 5.2: Automated RAG triggers

- **Scores of 1 to 2** indicate a lack of compensatory capability by the RAG system, suggesting the need for maintenance or possibly model fine-tuning. RAG will be temporarily deactivated until the system is improved. The user input will be processed but there will be no retrieval.
- **Scores of 3 to 4** initiate an augmentation with human-expert feedback only, utilizing flashcards or snippets to refine the output. Document-based RAG will be deactivated, but the human-expert feedback data will augment the input.
- **Scores of 5** initiate keyword-search RAG enhanced by previously gathered HF when necessary, utilizing flashcards or targeted information snippets to refine the output. The user is not required to provide new feedback in this case.

This program implements one of many scenarios. The scoring system, score levels, and triggers will vary from one project to another, depending on the specification goals to attain. It is recommended to organize workshops with a panel of users to decide how to implement this adaptive RAG system.

This adaptive approach aims to optimize the balance between automated retrieval and human insight, ensuring the generative model's outputs are of the highest possible relevance and accuracy. Let's now enter the input.

2.2. Input

A user of Company C is prompted to enter a question:

```
# Request user input for keyword parsing
user_input = input("Enter your query: ").lower()
```

In this example and program, we will focus on one question and topic: What is an LLM?. The question appears and is memorized by the model:

```
Enter your query: What is an LLM?
```

This program is a proof of concept with a strategy and example for the panel of users in Company C who wish to understand an LLM. Other topics can be added, and the program can be expanded to meet further needs. It is recommended to organize workshops with a panel of users to decide the next steps.

We have prepared the environment and will now activate a RAG scenario.

2.3. Mean ranking simulation scenario

For the sake of this program, let's assume that the human user feedback panel has been evaluating the hybrid adaptive RAG system for some time with the functions provided in sections *3.2. Human user rating* and *3.3. Human-expert evaluation*. The user feedback panel ranks the responses a number of times, which automatically updates by calculating the mean of the ratings and storing it in a ranking variable named ranking. The ranking score will help the management team decide whether to downgrade the rank of a document, upgrade it, or suppress documents through manual or automated functions. You can even simulate one of the scenarios described in the section *2.1. Integrating HF-RAG for augmented document inputs*.

We will begin with a 1 to 5 ranking, which will deactivate RAG so that we can see the native response of the generative model:

```
#Select a score between 1 and 5 to run the simulation
ranking=1
```

Then, we will modify this value to activate RAG without additional human-expert feedback with `ranking=5`. Finally, we will modify this value to activate human feedback RAG without retrieving documents with `ranking=3`.

In a real-life environment, these rankings will be triggered automatically with the functionality described in sections *3.2* and *3.3* after user feedback panel workshops are organized to define the system's expected behavior. If you wish to run the three scenarios described in section *2.1*, make sure to initialize the `text_input` variable that the generative model processes to respond:

```
# initializing the text for the generative AI model simulations
text_input=[]
```

Each time you switch scenarios, make sure to come back and reinitialize `text_input`.

 Due to its probabilistic nature, the generative AI model's output may vary from one run to another.

Let's go through the three rating categories described in section *2.1*.

Ranking 1–2: No RAG

The ranking of the generative AI's output is very low. All RAG functionality is deactivated until the management team can analyze and improve the system. In this case, `text_input` is equal to `user_input`:

```
if ranking>=1 and ranking<3:
    text_input=user_input
```

The generative AI model, in this case, GPT-4o, will generate the following output in section *2.6. Content generation*:

```
GPT-4 Response:
---------------
It seems like you're asking about "LLM" which stands for "Language Model for
Dialogue Applications" or more commonly referred to as a "Large Language
Model."
An LLM is a type of artificial intelligence model designed to understand,
generate, and interact with human language. These models are trained on vast
amounts of text data and use this training to generate text, answer questions,
summarize information, translate languages, and perform other language-related
tasks. They are a subset of machine learning models known as transformers,
which have been revolutionary in the field of natural language processing
(NLP).
```

```
Examples of LLMs include OpenAI's GPT (Generative Pre-trained Transformer)
series and Google's BERT (Bidirectional Encoder Representations from
Transformers).
---------------
```

This output cannot satisfy the user panel of Company C in this particular use case. They cannot relate this explanation to their customer service issues. Furthermore, many users will not bother going further since they have described their needs to the management team and expect pertinent responses. Let's see what human-expert feedback RAG can provide.

Ranking 3—4: Human-expert feedback RAG

In this scenario, human-expert feedback (see *section 3.4. Human-expert evaluation*) was triggered by poor user feedback ratings with automated RAG documents (ranking=5) and without RAG (ranking 1-2). The human-expert panel has filled in a flashcard, which has now been stored as an expert-level RAG document.

The program first checks the ranking and activates HF retrieval:

```
hf=False
if ranking>3 and ranking<5:
  hf=True
```

The program will then fetch the proper document from an expert panel (selected experts within a corporation) dataset based on keywords, embeddings, or other search methods that fit the goals of a project. In this case, we assume we have found the right flashcard and download it:

```
if hf==True:
  from grequests import download
  directory = "Chapter05"
  filename = "human_feedback.txt"
  download(directory, filename, private_token)
```

We verify if the file exists and load its content, clean it, store it in content, and assign it to text_input for the GPT-4 model:

```
if hf==True:
  # Check if 'human_feedback.txt' exists
    efile = os.path.exists('human_feedback.txt')

    if efile:
        # Read and clean the file content
        with open('human_feedback.txt', 'r') as file:
            content = file.read().replace('\n', ' ').replace('#', '')  #
Removing new line and markdown characters
            #print(content)  # Uncomment for debugging or maintenance display
        text_input=content
```

```
        print(text_input)
    else:
      print("File not found")
      hf=False
```

The content of the file explains both what an LLM is and how it can help Company C improve customer support:

```
A Large Language Model (LLM) is an advanced AI system trained on vast
amounts of text data to generate human-like text responses. It understands
and generates language based on the patterns and information it has learned
during training. LLMs are highly effective in various language-based tasks,
including answering questions, making recommendations, and facilitating
conversations. They can be continually updated with new information and trained
to understand specific domains or industries.For the C-phone series customer
support, incorporating an LLM could significantly enhance service quality and
efficiency. The conversational agent powered by an LLM can provide instant
responses to customer inquiries, reducing wait times and freeing up human
agents for more complex issues. It can be programmed to handle common technical
questions about the C-phone series, troubleshoot problems, guide users
through setup processes, and offer tips for optimizing device performance.
Additionally, it can be used to gather customer feedback, providing valuable
insights into user experiences and product performance. This feedback can then
be used to improve products and services. Furthermore, the LLM can be designed
to escalate issues to human agents when necessary, ensuring that customers
receive the best possible support at all levels. The agent can also provide
personalized recommendations for customers based on their usage patterns and
preferences, enhancing user satisfaction and loyalty.
```

If you now run sections *2.4* and *2.5* once and section *2.6* to generate the content based on this text_ input, the response will be satisfactory:

```
GPT-4 Response:
---------------

A Large Language Model (LLM) is a sophisticated AI system trained on extensive
text data to generate human-like text responses. It understands and generates
language based on patterns and information learned during training. LLMs are
highly effective in various language-based tasks such as answering questions,
making recommendations, and facilitating conversations. They can be
continuously updated with new information and trained to understand specific
domains or industries.  For the C-phone series customer support, incorporating
an LLM could significantly enhance service quality and efficiency. The
conversational agent powered by an LLM can provide instant responses to
customer inquiries, reducing wait times and freeing up human agents for more
complex issues.
```

```
It can be programmed to handle common technical questions about the C-phone
series,
troubleshoot problems, guide users through setup processes, and offer tips for
optimizing device performance. Additionally, it can be used to gather customer
feedback, providing valuable insights into user experiences and product
performance. This feedback can then be used to improve products and services.
Furthermore, the LLM can be designed to escalate issues to human agents when
necessary, ensuring that customers receive the best possible support at all
levels. The agent can also provide personalized recommendations for customers
based on their usage patterns and preferences, enhancing user satisfaction and
loyalty.
----------------
```

The preceding response is now much better since it defines LLMs and also shows how to improve customer service for Company C's C-phone series.

We will take this further in *Chapter 9, Empowering AI Models: Fine-Tuning RAG Data and Human Feedback*, in which we will fine-tune a generative model daily (or as frequently as possible) to improve its responses, thus alleviating the volume of RAG data. But for now, let's see what the system can achieve without HF but with RAG documents.

Ranking 5: RAG with no human-expert feedback documents

Some users do not require RAG documents that include human-expert RAG flashcards, snippets, or documents. This might be the case, particularly, if software engineers are the users.

In this case, the maximum number of words is limited to 100 to optimize API costs, but can be modified as you wish using the following code:

```python
if ranking>=5:
    max_words=100 #Limit: the size of the data we can add to the input
    rdata=process_query(user_input,max_words)
    print(rdata) # for maintenance if necessary
    if rdata:
        rdata_clean = rdata.replace('\n', ' ').replace('#', '')
        rdata_sentences = rdata_clean.split('. ')
        print(rdata)
    text_input=rdata
    print(text_input)
```

When we run the generative AI model, a reasonable output is produced that software engineers can relate to their business:

```
GPT-4 Response:
----------------
```

```
A large language model (LLM) is a type of language model known for its
capability to perform general-purpose language generation and other natural
language processing tasks such as classification. LLMs develop these
abilities by learning statistical relationships from text documents through a
computationally intensive training process that includes both self-supervised
and semi-supervised learning. These models can generate text, a form of
generative AI, by taking an input text and repeatedly predicting the next token
or word. LLMs are based on artificial neural networks. As of March 2024, the
most advanced and capable LLMs are constructed using a decoder-only transformer
architecture.
---------------
```

We can see that the output refers to March 2024 data, although GPT-4-turbo's training cutoff date was in December 2023, as explained in OpenAI's documentation: `https://platform.openai.com/docs/models/gpt-4-turbo-and-gpt-4`.

In production, at the end-user level, the error in the output can come from the data retrieved or the generative AI model. This shows the importance of HF. In this case, this error will hopefully be corrected in the retrieval documents or by the generative AI model. But we left the error in to illustrate that HF is not an option but a necessity.

These temporal RAG augmentations clearly justify the need for RAG-driven generative AI. However, it remains up to the users to decide if these types of outputs are sufficient or require more corporate customization in closed environments, such as within or for a company.

For the remainder of this program, let's assume `ranking>=5` for the next steps to show how the evaluator is implemented in the *Evaluator* section. Let's install the generative AI environment to generate content based on the user input and the document retrieved.

2.4.−2.5. Installing the generative AI environment

2.4. Checking the input before running the generator displays the user input and retrieved document before augmenting the input with this information. Then we continue to *2.5. Installing the generative AI environment*.

Only run this section once. If you modified the scenario in section *2.3*, you can skip this section to run the generative AI model again. This installation is not at the top of this notebook because a project team may choose to run this part of the program in another environment or even another server in production.

It is recommended to separate the retriever and generator functions as much as possible since they might be activated by different programs and possibly at different times. One development team might only work on the retriever functions while another team works on the generator functions.

We first install OpenAI:

```
!pip install openai==1.40.3
```

Then, we retrieve the API key. Store your OpenAI key in a safe location. In this case, it is stored on Google Drive:

```
#API Key
#Store your key in a file and read it(you can type it directly in the notebook
but it will be visible for somebody next to you)
from google.colab import drive
drive.mount('/content/drive')

f = open("drive/MyDrive/files/api_key.txt", "r")
API_KEY=f.readline().strip()
f.close()

#The OpenAI Key
import os
import openai
os.environ['OPENAI_API_KEY'] =API_KEY
openai.api_key = os.getenv("OPENAI_API_KEY")
```

We are now all set for content generation.

2.6. Content generation

To generate content, we first import and set up what we need. We've introduced time to measure the speed of the response and have chosen gpt-4o as our conversational model:

```
import openai
from openai import OpenAI
import time

client = OpenAI()
gptmodel="gpt-4o"
start_time = time.time()  # Start timing before the request
```

We then define a standard Gpt-4o prompt, giving it enough information to respond and leaving the rest up to the model and RAG data:

```
def call_gpt4_with_full_text(itext):
    # Join all lines to form a single string
    text_input = '\n'.join(itext)
    prompt = f"Please summarize or elaborate on the following content:\n{text_
input}"
```

```
    try:
        response = client.chat.completions.create(
            model=gptmodel,
            messages=[
                {"role": "system", "content": "You are an expert Natural Language
Processing exercise expert."},
                {"role": "assistant", "content": "1.You can explain read the input
and answer in detail"},
                {"role": "user", "content": prompt}
            ],
            temperature=0.1  # Add the temperature parameter here and other
parameters you need
        )
    return response.choices[0].message.content.strip()
    except Exception as e:
        return str(e)
```

The code then formats the output:

```
import textwrap


def print_formatted_response(response):
    # Define the width for wrapping the text
    wrapper = textwrap.TextWrapper(width=80)  # Set to 80 columns wide, but
adjust as needed
    wrapped_text = wrapper.fill(text=response)

    # Print the formatted response with a header and footer
    print("GPT-4 Response:")
    print("---------------")
    print(wrapped_text)
    print("--------------\n")


# Assuming 'gpt4_response' contains the response from the previous GPT-4 call
print_formatted_response(gpt4_response)
```

The response is satisfactory in this case, as we saw in section 2.3. In the ranking=5 scenario, which is the one we are now evaluating, we get the following output:

```
GPT-4 Response:
---------------
GPT-4 Response:
---------------
```

```
### Summary: A large language model (LLM) is a computational model known for
its ability to perform general-purpose language generation and other natural
language processing tasks, such as classification. LLMs acquire these abilities
by learning statistical relationships from vast amounts of text during a
computationally intensive self-supervised and semi-supervised training process.
They can be used for text generation, a form of generative AI, by taking input
text and repeatedly predicting the next token or word. LLMs are artificial
neural networks that use the transformer architecture...
```

The response looks fine, but is it really accurate? Let's run the evaluator to find out.

3. Evaluator

Depending on each project's specifications and needs, we can implement as many mathematical and human evaluation functions as necessary. In this section, we will implement two automatic metrics: response time and cosine similarity score. We will then implement two interactive evaluation functions: human user rating and human-expert evaluation.

3.1. Response time

The response time was calculated and displayed in the API call with:

```
import time

...

start_time = time.time()  # Start timing before the request

...

response_time = time.time() - start_time  # Measure response time
print(f"Response Time: {response_time:.2f} seconds")  # Print response time
```

In this case, we can display the response time without further development:

```
print(f"Response Time: {response_time:.2f} seconds")  # Print response time
```

The output will vary depending on internet connectivity and the capacity of OpenAI's servers. In this case, the output is:

```
Response Time: 7.88 seconds
```

It seems long, but online conversational agents take some time to answer as well. Deciding if this performance is sufficient remains a management decision. Let's run the cosine similarity score next.

3.2. Cosine similarity score

Cosine similarity measures the cosine of the angle between two non-zero vectors. In the context of text analysis, these vectors are typically **TF-IDF** (**Term Frequency-Inverse Document Frequency**) representations of the text, which weigh terms based on their importance relative to the document and a corpus.

GPT-4o's input, which is `text_input`, and the model's response, which is `gpt4_response`, are treated by TF-IDF as two separate "documents." The `vectorizer` transforms the documents into vectors. Then, vectorization considers how terms are shared and emphasized between the input and the response with the `vectorizer.fit_transform([text1, text2])`.

The goal is to quantify the thematic and lexical overlap through the following function:

```python
from sklearn.feature_extraction.text import TfidfVectorizer
from sklearn.metrics.pairwise import cosine_similarity


def calculate_cosine_similarity(text1, text2):
    vectorizer = TfidfVectorizer()
    tfidf = vectorizer.fit_transform([text1, text2])
    similarity = cosine_similarity(tfidf[0:1], tfidf[1:2])
    return similarity[0][0]


# Example usage with your existing functions
similarity_score = calculate_cosine_similarity(text_input, gpt4_response)


print(f"Cosine Similarity Score: {similarity_score:.3f}")
```

Cosine similarity relies on `TfidfVectorizer` to transform the two documents into TF-IDF vectors. The `cosine_similarity` function then calculates the similarity between these vectors. A result of 1 indicates identical texts, while 0 shows no similarity. The output of the function is:

```
Cosine Similarity Score: 0.697
```

The score shows a strong similarity between the input and the output of the model. But how will a human user rate this response? Let's find out.

3.3. Human user rating

The human user rating interface provides human user feedback. As reiterated throughout this chapter, I recommend designing this interface and process after fully understanding user needs through a workshop with them. In this section, we will assume that the human user panel is a group of software developers testing the system.

The code begins with the interface's parameters:

```python
# Score parameters
counter=20                      # number of feedback queries
score_history=30                # human feedback
threshold=4                     # minimum rankings to trigger human expert
feedback
```

In this simulation, the parameters show that the system has computed human feedback:

- `counter=20` shows the number of ratings already entered by the users
- `score_history=60` shows the total score of the 20 ratings
- `threshold=4` states the minimum mean rating, `score_history/counter`, to obtain without triggering a human-expert feedback request

We will now run the interface to add an instance to these parameters. The provided Python code defines the `evaluate_response` function, designed to assess the relevance and coherence of responses generated by a language model such as GPT-4. Users rate the generated text on a scale from 1 (poor) to 5 (excellent), with the function ensuring valid input through recursive checks. The code calculates statistical metrics like mean scores to gauge the model's performance over multiple evaluations.

The evaluation function is a straightforward feedback request to obtain values between 1 and 5:

```python
import numpy as np
def evaluate_response(response):
    print("\nGenerated Response:")
    print(response)
    print("\nPlease evaluate the response based on the following criteria:")
    print("1 - Poor, 2 - Fair, 3 - Good, 4 - Very Good, 5 - Excellent")
    score = input("Enter the relevance and coherence score (1-5): ")
    try:
        score = int(score)
        if 1 <= score <= 5:
            return score
        else:
            print("Invalid score. Please enter a number between 1 and 5.")
            return evaluate_response(response)  # Recursive call if the input
is invalid
    except ValueError:
        print("Invalid input. Please enter a numerical value.")
        return evaluate_response(response)  # Recursive call if the input is
invalid
```

We then call the function:

```python
score = evaluate_response(gpt4_response)
print("Evaluator Score:", score)
```

The function first displays the response, as shown in the following excerpt:

```
Generated Response:
### Summary:
A large language model (LLM) is a computational model…
```

Then, the user enters an evaluation score between 1 and 5, which is 1 in this case:

```
Please evaluate the response based on the following criteria:
1 - Poor, 2 - Fair, 3 - Good, 4 - Very Good, 5 - Excellent
Enter the relevance and coherence score (1-5): 3
```

The code then computes the statistics:

```
counter+=1
score_history+=score
mean_score=round(np.mean(score_history/counter), 2)
if counter>0:
  print("Rankings      :", counter)
  print("Score history : ", mean_score)
```

The output shows a relatively very low rating:

```
Evaluator Score: 3
Rankings      : 21
Score history :  3.0
```

The evaluator score is 3, the overall ranking is 3, and the score history is 3 also! Yet, the cosine similarity was positive. The human-expert evaluation request will be triggered because we set the threshold to 4:

```
threshold=4
```

What's going on? Let's ask an expert and find out!

3.4. Human-expert evaluation

Metrics such as cosine similarity indeed measure similarity but not in-depth accuracy. Time performance will not determine the accuracy of a response either. But if the rating is too low, why is that? Because the user is not satisfied with the response!

The code first downloads thumbs-up and thumbs-down images for the human-expert user:

```
from grequests import download

# Define your variables
directory = "commons"
filename = "thumbs_up.png"
download(directory, filename, private_token)

# Define your variables
directory = "commons"
filename = "thumbs_down.png"
download(directory, filename, private_token)
```

The parameters to trigger an expert's feedback are `counter_threshold` and `score_threshold`. The number of user ratings must exceed the expert's threshold counter, which is `counter_threshold=10`. The threshold of the mean score of the ratings is 4 in this scenario: `score_threshold=4`. We can now simulate the triggering of an expert feedback request:

```
if counter>counter_threshold and score_history<=score_threshold:
    print("Human expert evaluation is required for the feedback loop.")
```

In this case, the output will confirm the expert feedback loop because of the poor mean ratings and the number of times the users rated the response:

```
Human expert evaluation is required for the feedback loop.
```

As mentioned, in a real-life project, a workshop with expert users should be organized to define the interface. In this case, a standard HTML interface in a Python cell will display the thumbs-up and thumbs-down icons. If the expert presses on the thumbs-down icon, a feedback snippet can be entered and saved in a feedback file named `expert_feedback.txt`, as shown in the following excerpt of the code:

```python
import base64
from google.colab import output
from IPython.display import display, HTML


def image_to_data_uri(file_path):
    """
    Convert an image to a data URI.
    """
    with open(file_path, 'rb') as image_file:
        encoded_string = base64.b64encode(image_file.read()).decode()
        return f'data:image/png;base64,{encoded_string}'

thumbs_up_data_uri = image_to_data_uri('/content/thumbs_up.png')
thumbs_down_data_uri = image_to_data_uri('/content/thumbs_down.png')


def display_icons():
    # Define the HTML content with the two clickable images
.../...
def save_feedback(feedback):
    with open('/content/expert_feedback.txt', 'w') as f:
        f.write(feedback)
    print("Feedback saved successfully.")


# Register the callback
output.register_callback('notebook.save_feedback', save_feedback)
```

```
print("Human Expert Adaptive RAG activated")

# Display the icons with click handlers
display_icons()
```

The code will display the icons shown in the following figure. If the expert user presses the thumbs-down icon, they will be prompted to enter feedback.

Figure 5.3: Feedback icons

You can add a function for thumbs-down meaning that the response was incorrect and that the management team has to communicate with the user panel or add a prompt to the user feedback interface. This is a management decision, of course. In our scenario, the human expert pressed the thumbs-down icon and was prompted to enter a response:

Figure 5.4: "Enter feedback" prompt

The human expert provided the response, which was saved in '/content/expert_feedback.txt'. Through this, we have finally discovered the inaccuracy, which is in the content of the file displayed in the following cell:

```
There is an inaccurate statement in the text:
"As of March 2024, the largest and most capable LLMs are built with a decoder-
only transformer-based architecture."

This statement is not accurate because the largest and most capable
Large Language Models, such as Meta's Llama models, have a transformer-
based architecture, but they are not "decoder-only." These models use the
architecture of the transformer, which includes both encoder and decoder
components.
```

The preceding expert's feedback can then be used to improve the RAG dataset. With this, we have explored the depths of HF-RAG interactions. Let's summarize our journey and move on to the next steps.

Summary

As we wrap up our hands-on approach to pragmatic AI implementations, it's worth reflecting on the transformative journey we've embarked on together, exploring the dynamic world of adaptive RAG. We first examined how HF not only complements but also critically enhances generative AI, making it a more powerful tool customized to real-world needs. We described the adaptive RAG ecosystem and then went hands-on, building from the ground up. Starting with data collection, processing, and querying, we integrated these elements into a RAG-driven generative AI system. Our approach wasn't just about coding; it was about adding adaptability to AI through continuous HF loops.

By augmenting GPT-4's capabilities with expert insights from previous sessions and end-user evaluations, we demonstrated the practical application and significant impact of HF. We implemented a system where the output is not only generated but also ranked by end-users. Low rankings triggered an expert feedback loop, emphasizing the importance of human intervention in refining AI responses. Building an adaptive RAG program from scratch ensured a deep understanding of how integrating HF can shift a standard AI system to one that evolves and improves over time.

This chapter wasn't just about learning; it was about doing, reflecting, and transforming theoretical knowledge into practical skills. We are now ready to scale RAG-driven AI to production-level volumes and complexity in the next chapter.

Questions

Answer the following questions with *Yes* or *No*:

1. Is human feedback essential in improving RAG-driven generative AI systems?
2. Can the core data in a generative AI model be changed without retraining the model?
3. Does Adaptive RAG involve real-time human feedback loops to improve retrieval?
4. Is the primary focus of Adaptive RAG to replace all human input with automated responses?
5. Can human feedback in Adaptive RAG trigger changes in the retrieved documents?
6. Does Company C use Adaptive RAG solely for customer support issues?
7. Is human feedback used only when the AI responses have high user ratings?
8. Does the program in this chapter provide only text-based retrieval outputs?
9. Is the Hybrid Adaptive RAG system static, meaning it cannot adjust based on feedback?
10. Are user rankings completely ignored in determining the relevance of AI responses?

References

- *Studying Large Language Model Behaviors Under Realistic Knowledge Conflicts* by Evgenii Kortukov, Alexander Rubinstein, Elisa Nguyen, Seong Joon Oh: https://arxiv.org/abs/2404.16032
- OpenAI models: https://platform.openai.com/docs/models

Further reading

For more information on the vectorizer and cosine similarity functionality implemented in this chapter, use the following links:

- Feature extraction – TfidfVectorizer: https://scikit-learn.org/stable/modules/generated/sklearn.feature_extraction.text.TfidfVectorizer.html
- sklearn.metrics – cosine_similarity: https://scikit-learn.org/stable/modules/generated/sklearn.metrics.pairwise.cosine_similarity.html

Learn further in a live workshop with the author

If you are curious how RAG-based systems scale into agentic workflows using context engineering, consider registering for the author's live workshop on 24 January 2026, which extends the ideas discussed here into hands-on practice.

https://packt.link/gk5Yu

6

Scaling RAG Bank Customer Data with Pinecone

Scaling up RAG documents, whether text-based or multimodal, isn't just about piling on and accumulating more data—it fundamentally changes how an application works. Firstly, scaling is about finding the right amount of data, not just more of it. Secondly, as you add more data, the demands on an application can change—it might need new features to handle the bigger load. Finally, cost monitoring and speed performance will constrain our projects when scaling. Hence, this chapter is designed to equip you with cutting-edge techniques for leveraging AI in solving the real-world scaling challenges you may face in your projects. For this, we will be building a recommendation system based on pattern-matching using Pinecone to minimize bank customer churn (customers choosing to leave a bank).

We will start with a step-by-step approach to developing the first program of our pipeline. Here, you will learn how to download a Kaggle bank customer dataset and perform **exploratory data analysis (EDA)**. This foundational step is crucial as it guides and supports you in preparing your dataset and your RAG strategy for the next stages of processing. The second program of our pipeline introduces you to the powerful combination of Pinecone—a vector database suited for handling large-scale vector search—and OpenAI's `text-embedding-3-small` model. Here, you'll chunk and embed your data before upserting (updating or inserting records) it into a Pinecone index that we will scale up to 1,000,000+ vectors. We will ready it for complex query retrieval at a satisfactory speed. Finally, the third program of our pipeline will show you how to build RAG queries using Pinecone, augment user input, and leverage GPT-4o to generate AI-driven recommendations. The goal is to reduce churn in banking by offering personalized, insightful recommendations. By the end of this chapter, you'll have a good understanding of how to apply the power of Pinecone and OpenAI technologies to your RAG projects.

To sum up, this chapter covers the following topics:

- The key aspects of scaling RAG vector stores
- EDA for data preparation
- Scaling with Pinecone vector storage
- Chunking strategy for customer bank information
- Embedding data with OpenAI embedding models
- Upserting data
- Using Pinecone for RAG
- Generative AI-driven recommendations with GPT-4o to reduce bank customer churn

Let's begin by defining how we will scale with Pinecone.

Scaling with Pinecone

We will be implementing Pinecone's innovative vector database technology with OpenAI's powerful embedding capabilities to construct data processing and querying systems. The goal is to build a recommendation system to encourage customers to continue their association with a bank. Once you understand this approach, you will be able to apply it to any domain requiring recommendations (leisure, medical, or legal). To understand and optimize the complex processes involved, we will build the programs from scratch with a minimal number of components. In this chapter, we will use the Pinecone vector database and the OpenAI LLM model.

Selecting and designing an architecture depends on a project's specific goals. Depending on your project's needs, you can apply this methodology to other platforms. In this chapter and architecture, the combination of a vector store and a generative AI model is designed to streamline operations and facilitate scalability. With that context in place, let's go through the architecture we will be building in Python.

Architecture

In this chapter, we will implement vector-based similarity search functionality, as we did in *Chapter 2, RAG Embedding Vector Stores with Deep Lake and OpenAI*, and *Chapter 3, Building Index-Based RAG with LlamaIndex, Deep Lake, and OpenAI*. We will take the structure of the three pipelines we designed in those chapters and apply them to our recommendation system, as shown in *Figure 6.1*. If necessary, take the time to go through those chapters before implementing the code in this chapter.

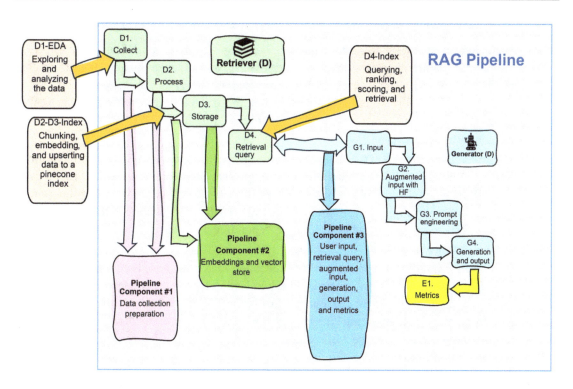

Figure 6.1: Scaling RAG-driven generative AI pipelines

The key features of the scaled recommendation system we will build can be summarized in the three pipelines shown in the preceding figure:

- **Pipeline 1: Collecting and preparing the dataset**

    In this pipeline, we will perform EDA on the dataset with standard queries and k-means clustering.

- **Pipeline 2: Scaling a Pinecone index (vector store)**

    In this pipeline, we will see how to chunk, embed, and upsert 1,000,000+ documents to a Pinecone index (vector store).

- **Pipeline 3: RAG generative AI**

    This pipeline will take us to fully scaled RAG when we query a 1,000,000+ vector store and augment the input of a GPT-4o model to make targeted recommendations.

The main theoretical and practical applications of the three programs we will explore include:

- **Scalable and serverless infrastructure**: We begin by understanding Pinecone's serverless architecture, which eliminates the complexities of server management and scaling. We don't need to manage storage resources or machine usage. It's a pay-as-you-go approach based on serverless indexes formed by a cloud and region, for example, **Amazon Web Services** (**AWS**) in us-east-1. Scaling and billing are thus simplified, although we still have to monitor and minimize the costs!

- **Lightweight and simplified development environment**: Our integration strategy will minimize the use of external libraries, maintaining a lightweight development stack. Directly using OpenAI to generate embeddings and Pinecone to store and query these embeddings simplifies the data processing pipeline and increases system efficiency. Although this approach can prove effective, other methods are possible depending on your project, as implemented in other chapters of this book.

- **Optimized scalability and performance**: Pinecone's vector database is engineered to handle large-scale datasets effectively, ensuring that application performance remains satisfactory as the data volume grows. As for all cloud platforms and APIs, examine the privacy and security constraints when implementing Pinecone and OpenAI. Also, continually monitor the system's performance and costs, as we will see in the *Pipeline 2: Scaling a Pinecone index (vector store)* section of this chapter.

Let's now go to our keyboards to collect and process the Bank Customer Churn dataset.

Pipeline 1: Collecting and preparing the dataset

This section will focus on handling and analyzing the Bank Customer Churn dataset. We will guide you through the steps of setting up your environment, manipulating data, and applying **machine learning (ML)** techniques. It is important to get the "feel" of a dataset with human analysis before using algorithms as tools. Human insights will always remain critical because of the flexibility of human creativity. As such, we will implement data collection and preparation in Python in three main steps:

1. **Collecting and processing the dataset:**

    - Setting up the Kaggle environment to authenticate and download datasets
    - Collecting and unzipping the Bank Customer Churn dataset
    - Simplifying the dataset by removing unnecessary columns

2. **Exploratory data analysis:**

    - Performing initial data inspections to understand the structure and type of data we have
    - Investigating relationships between customer complaints and churn (closing accounts)
    - Exploring how age and salary levels relate to customer churn
    - Generating a heatmap to visualize correlations between numerical features

3. **Training an ML model:**

- Preparing the data for ML
- Applying clustering techniques to discover patterns in customer behavior
- Assessing the effectiveness of different cluster configurations
- Concluding and moving on to RAG-driven generative AI

Our goal is to analyze the dataset and prepare it for *Pipeline 2: Scaling a Pinecone index (vector store)*. To achieve that goal, we need to perform a preliminary EDA of the dataset. Moreover, each section is designed to be a hands-on walkthrough of the code from scratch, ensuring you gain practical experience and insights into data science workflows. Let's get started by collecting the dataset.

1. Collecting and processing the dataset

Let's first collect the Bank Customer Churn dataset on Kaggle and process it:

`https://www.kaggle.com/datasets/radheshyamkollipara/bank-customer-churn`

The file `Customer-Churn-Records.csv` contains data on 10,000 records of customers from a bank focusing on various aspects that might influence customer churn. The dataset was uploaded by Radheshyam Kollipara, who rightly states:

As we know, it is much more expensive to sign in a new client than keeping an existing one. It is advantageous for banks to know what leads a client towards the decision to leave the company. Churn prevention allows companies to develop loyalty programs and retention campaigns to keep as many customers as possible.

Here are the details of the columns included in the dataset that follow the description on Kaggle:

`RowNumber`—corresponds to the record (row) number and has no effect on the output.

`CustomerId`—contains random values and has no effect on customers leaving the bank.

`Surname`—the surname of a customer has no impact on their decision to leave the bank.

`CreditScore`—can have an effect on customer churn since a customer with a higher credit score is less likely to leave the bank.

`Geography`—a customer's location can affect their decision to leave the bank.

`Gender`—it's interesting to explore whether gender plays a role in a customer leaving the bank.

`Age`—this is certainly relevant since older customers are less likely to leave their bank than younger ones.

Tenure—refers to the number of years that the customer has been a client of the bank. Normally, older clients are more loyal and less likely to leave a bank.

Balance—is also a very good indicator of customer churn, as people with a higher balance in their accounts are less likely to leave the bank compared to those with lower balances.

NumOfProducts—refers to the number of products that a customer has purchased through the bank.

HasCrCard—denotes whether or not a customer has a credit card. This column is also relevant since people with a credit card are less likely to leave the bank.

IsActiveMember—active customers are less likely to leave the bank.

EstimatedSalary—as with balance, people with lower salaries are more likely to leave the bank compared to those with higher salaries.

Exited—whether or not the customer left the bank.

Complain—customer has complained or not.

Satisfaction Score—Score provided by the customer for their complaint resolution.

Card Type—the type of card held by the customer.

Points Earned—the points earned by the customer for using a credit card.

Now that we know what the dataset contains, we need to collect it and process it for EDA. Let's install the environment.

Installing the environment for Kaggle

To collect datasets from Kaggle automatically, you will need to sign up and create an API key at https://www.kaggle.com/. At the time of writing this notebook, downloading datasets is free. Follow the instructions to save and use your Kaggle API key. Store your key in a safe location. In this case, the key is in a file on Google Drive that we need to mount:

```
#API Key
#Store your key in a file and read it(you can type it directly in the notebook
but it will be visible for somebody next to you)
from google.colab import drive
drive.mount('/content/drive')
```

The program now reads the JSON file and sets environment variables for Kaggle authentication using your username and an API key:

```
import os
import json
```

```
with open(os.path.expanduser("drive/MyDrive/files/kaggle.json"), "r") as f:
    kaggle_credentials = json.load(f)
kaggle_username = kaggle_credentials["username"]
kaggle_key = kaggle_credentials["key"]

os.environ["KAGGLE_USERNAME"] = kaggle_username
os.environ["KAGGLE_KEY"] = kaggle_key
```

We are now ready to install Kaggle and authenticate it:

```
try:
  import kaggle
except:
  !pip install kaggle
import kaggle
kaggle.api.authenticate()
```

And that's it! That's all we need. We are now ready to collect the Bank Customer Churn dataset.

Collecting the dataset

We will now download the zipped dataset, extract the CSV file, upload it into a pandas DataFrame, drop columns that we will not use, and display the result. Let's first download the zipped dataset:

```
!kaggle datasets download -d radheshyamkollipara/bank-customer-churn
```

The output displays the source of the data:

```
Dataset URL: https://www.kaggle.com/datasets/radheshyamkollipara/bank-customer-
churn
License(s): other
bank-customer-churn.zip: Skipping, found more recently modified local copy (use
--force to force download)
```

We can now unzip the data:

```
import zipfile

with zipfile.ZipFile('/content/bank-customer-churn.zip', 'r') as zip_ref:
    zip_ref.extractall('/content/')

print("File Unzipped!")
```

The output should confirm that the file is unzipped:

```
File Unzipped!
```

The CSV file is uploaded to a pandas DataFrame named data1:

```
import pandas as pd
# Load the CSV file
file_path = '/content/Customer-Churn-Records.csv'
data1 = pd.read_csv(file_path)
```

We will now drop the following four columns in this scenario:

- RowNumber: We don't need these columns because we will be creating a unique index for each record.

- Surname: The goal in this scenario is to anonymize the data and not display surnames. We will focus on customer profiles and behaviors, such as complaints and credit card consumption (points earned).

- Gender: Consumer perceptions and behavior have evolved in the 2020s. It is more ethical and just as efficient to leave this information out in the context of a sample project.

- Geography: This field might be interesting in some cases. For this scenario, let's leave this feature out to avoid overfitting outputs based on cultural clichés. Furthermore, including this feature would require more information if we wanted to calculate distances for delivery services, for example:

```
# Drop columns and update the DataFrame in place
data1.drop(columns=['RowNumber','Surname', 'Gender','Geography'], inplace=True)
data1
```

The output triggered by data1 shows a simplified yet sufficient dataset:

	CustomerId	CreditScore	Age	Tenure	Balance	NumOfProducts	HasCrCard	IsActiveMember	EstimatedSalary
0	15634602	619	42	2	0.00	1	1	1	101348.88
1	15647311	608	41	1	83807.86	1	0	1	112542.58
2	15619304	502	42	8	159660.80	3	1	0	113931.57
3	15701354	699	39	1	0.00	2	0	0	93826.63
4	15737888	850	43	2	125510.82	1	1	1	79084.10
...	...	...	...	...	...	...	...	...	...
9995	15606229	771	39	5	0.00	2	1	0	96270.64
9996	15569892	516	35	10	57369.61	1	1	1	101699.77
9997	15584532	709	36	7	0.00	1	0	1	42085.58
9998	15682355	772	42	3	75075.31	2	1	0	92888.52
9999	15628319	792	28	4	130142.79	1	1	0	38190.78

0000 rows × 14 columns

Figure 6.2: Triggered output

This approach's advantage is that it optimizes the size of the data that will be inserted into the Pinecone index (vector store). Optimizing the data size before inserting data into Pinecone and reducing the dataset by removing unnecessary fields can be very beneficial. It reduces the amount of data that needs to be transferred, stored, and processed in the vector store. When scaling, smaller data sizes can lead to faster query performance and lower costs, as Pinecone pricing can depend on the amount of data stored and the computational resources used for queries.

We can now save the new pandas DataFrame in a safe location:

```
data1.to_csv('data1.csv', index=False)
!cp /content/data1.csv /content/drive/MyDrive/files/rag_c6/data1.csv
```

You can save it in the location that is best for you. Just make sure to save it because we will use it in the *Pipeline 2: Scaling a Pinecone index (vector store)* section of this chapter. We will now explore the optimized dataset before deciding how to implement it in a vector store.

2. Exploratory data analysis

In this section, we will perform EDA using the data that pandas has just defined, which contains customer data from a bank. EDA is a critical step before applying any RAG techniques with vector stores, as it helps us understand the underlying patterns and trends within the data.

For instance, our preliminary analysis shows a direct correlation between customer complaints and churn rates, indicating that customers who have lodged complaints are more likely to leave the bank. Additionally, our data reveals that customers aged 50 and above are less likely to churn compared to younger customers. Interestingly, income levels (particularly the threshold of $100,000) do not appear to significantly influence churn decisions.

Through the careful examination of these insights, we'll demonstrate why jumping straight into complex ML models, especially deep learning, may not always be necessary or efficient for drawing basic conclusions. In scenarios where the relationships within the data are evident and the patterns straightforward, simpler statistical methods or even basic data analysis techniques might be more appropriate and resource-efficient. For example, k-means clustering can be effective, and we will implement it in the *Training an ML model* section of this chapter.

However, this is not to understate the power of advanced RAG techniques, which we will explore in the *Pipeline 2: Scaling a Pinecone index (vector store)* section of this chapter. In that section, we will employ deep learning within vector stores to uncover more subtle patterns and intricate relationships that are not readily apparent through classic EDA.

If we display the columns of the DataFrame, we can see that it is challenging to find patterns:

```
 #   Column              Non-Null Count  Dtype
---  ------              --------------  -----
 0   CustomerId          10000 non-null  int64
 1   CreditScore         10000 non-null  int64
 2   Age                 10000 non-null  int64
 3   Tenure              10000 non-null  int64
```

```
4    Balance            10000 non-null  float64
5    NumOfProducts      10000 non-null  int64
6    HasCrCard          10000 non-null  int64
7    IsActiveMember     10000 non-null  int64
8    EstimatedSalary    10000 non-null  float64
9    Exited             10000 non-null  int64
10   Complain           10000 non-null  int64
11   Satisfaction Score 10000 non-null  int64
12   Card Type          10000 non-null  object
13   Point Earned       10000 non-null  int64
```

Age, EstimatedSalary, and Complain are possible determining features that could be correlated with Exited. We can also display the DataFrame to gain insights, as shown in the excerpt of data1 in the following figure:

Age	Tenure	Balance	NumOfProducts	HasCrCard	IsActiveMember	EstimatedSalary	Exited	Complain
42	2	0.00	1	1	1	101348.88	1	1
41	1	83807.86	1	0	1	112542.58	0	1
42	8	159660.80	3	1	0	113931.57	1	1
39	1	0.00	2	0	0	93826.63	0	0
43	2	125510.82	1	1	1	79084.10	0	0
...	...	...	...	...	...	...	...	...
39	5	0.00	2	1	0	96270.64	0	0
35	10	57369.61	1	1	1	101699.77	0	0
36	7	0.00	1	0	1	42085.58	1	1
42	3	75075.31	2	1	0	92888.52	1	1
28	4	130142.79	1	1	0	38190.78	0	0

Figure 6.3: Visualizing the strong correlation between customer complaints and bank churning (Exited)

The main feature seems to be Complain, which leads to Exited (churn), as shown by running a standard calculation on the DataFrame:

```
# Calculate the percentage of complain over exited
if sum_exited > 0:  # To avoid division by zero
    percentage_complain_over_exited = (sum_complain/ sum_exited) * 100
else:
    percentage_complain_over_exited = 0

# Print results
```

```
print(f"Sum of Exited = {sum_exited}")
print(f"Sum of Complain = {sum_complain}")
print(f"Percentage of complain over exited = {percentage_complain_over_
exited:.2f}%")
```

The output shows a very high 100.29% ratio between complaints and customers leaving the bank (churning). This means that customers who complained did in fact leave the bank, which is a natural market trend:

```
Sum of Exited = 2038
Sum of Complain = 2044
Percentage of complain over exited = 100.29%
```

We can see that only a few exited the bank (six customers) without complaining.

Run the following cells from GitHub; these contain Python functions that are variations of the exited and complain ratios and will produce the following outputs:

- Age and Exited with a threshold of age=50 shows that persons over 50 seem less likely to leave a bank:

```
Sum of Age 50 and Over among Exited = 634
Sum of Exited = 2038
Percentage of Age 50 and Over among Exited = 31.11%
```

  Conversely, the output shows that younger customers seem more likely to leave a bank if they are dissatisfied. You can explore different age thresholds to analyze the dataset further.

- Salary and Exited with a threshold of salary_threshold=100000 doesn't seem to be a significant feature, as shown in this output:

```
Sum of Estimated Salary over 100000 among Exited = 1045
Sum of Exited = 2038
Percentage of Estimated Salary over 100000 among Exited = 51.28%
```

  Try exploring different thresholds to analyze the dataset to confirm or refute this trend.

Let's create a heatmap based on the data1 pandas DataFrame:

```
import seaborn as sns
import matplotlib.pyplot as plt
# Select only numerical columns for the correlation heatmap
numerical_columns = data1.select_dtypes(include=['float64', 'int64']).columns
# Correlation heatmap
plt.figure(figsize=(12, 8))
sns.heatmap(data1[numerical_columns].corr(), annot=True, fmt='.2f',
cmap='coolwarm')
plt.title('Correlation Heatmap')
plt.show()
```

We can see that the highest correlation is between `Complain` and `Exited`:

Figure 6.4: Excerpt of the heatmap

The preceding heatmap visualizes the correlation between each pair of features (variables) in the dataset. It shows the correlation coefficients between each pair of variables, which can range from -1(low correlation) to 1(high correlation), with 0 indicating no correlation.

With that, we have explored several features. Let's build an ML model to take this exploration further.

3. Training an ML model

Let's continue our EDA and drill into the dataset further with an ML model. This section implements the training of an ML model using clustering techniques, specifically k-means clustering, to explore patterns within our dataset. We'll prepare and process data for analysis, apply clustering, and then evaluate the results using different metrics. This approach is valuable for extracting insights without immediately resorting to more complex deep learning methods.

> k-means clustering is an unsupervised ML algorithm that partitions a dataset into k distinct, non-overlapping clusters by minimizing the variance within each cluster. The algorithm iteratively assigns data points to one of the k clusters based on the nearest mean (centroid), which is recalculated after each iteration until convergence.

Now, let's break down the code section by section.

Data preparation and clustering

We will first copy our chapter's dataset `data1` to `data2` to be able to go back to `data1` if necessary if we wish to try other ML models:

```
# Copying data1 to data2
data2 = data1.copy()
```

You can explore the data with various scenarios of feature sets. In this case, we will select `'CreditScore'`, `'Age'`, `'EstimatedSalary'`, `'Exited'`, `'Complain'`, and `'Point Earned'`:

```
# Import necessary libraries
import pandas as pd
from sklearn.cluster import KMeans
from sklearn.preprocessing import StandardScaler
from sklearn.metrics import silhouette_score , davies_bouldin_score
# Assuming you have a dataframe named data1 loaded as described
# Selecting relevant features
features = data2[['CreditScore', 'Age', 'EstimatedSalary', 'Exited',
'Complain', 'Point Earned']]
```

As in standard practice, let's scale the features before running an ML model:

```
# Standardize the features
scaler = StandardScaler()
features_scaled = scaler.fit_transform(features)
```

The credit score, estimated salary, and points earned (reflecting credit card spending) are good indicators of a customer's financial standing with the bank. The age factor, combined with these other factors, might influence older customers to remain with the bank. However, the important point to note is that complaints may lead any market segment to consider leaving since complaints and churn are strongly correlated.

We will now try to find two to four clusters to find the optimal number of clusters for this set of features:

```
# Experiment with different numbers of clusters
for n_clusters in range(2, 5):  # Example range from 2 to 5
    kmeans = KMeans(n_clusters=n_clusters, n_init=20, random_state=0)
    cluster_labels = kmeans.fit_predict(features_scaled)
    silhouette_avg = silhouette_score(features_scaled, cluster_labels)
    db_index = davies_bouldin_score(features_scaled, cluster_labels)
    print(f'For n_clusters={n_clusters}, the silhouette score is {silhouette_
avg:.4f} and the Davies-Bouldin Index is {db_index:.4f}')
```

The output contains an evaluation of clustering performance using two metrics—the silhouette score and the Davies-Bouldin index—across different numbers of clusters (ranging from 2 to 4):

```
For n_clusters=2, the silhouette score is 0.6129 and the Davies-Bouldin Index
is 0.6144
For n_clusters=3, the silhouette score is 0.3391 and the Davies-Bouldin Index
is 1.1511
For n_clusters=4, the silhouette score is 0.3243 and the Davies-Bouldin Index
is 1.0802
```

Silhouette score: This metric measures the quality of clustering by calculating the mean intra-cluster distance (how close each point in one cluster is to points in the same cluster) and the mean nearest cluster distance (how close each point is to points in the next nearest cluster). The score ranges from -1 to 1, where a high value indicates that clusters are well-separated and internally cohesive. In this output, the highest silhouette score is 0.6129 for 2 clusters, suggesting better cluster separation and cohesion compared to 3 or 4 clusters.

Davies-Bouldin index: This index evaluates clustering quality by comparing the ratio of within-cluster distances to between-cluster distances. Lower values of this index indicate better clustering, as they suggest lower intra-cluster variance and higher separation between clusters. The smallest Davies-Bouldin index in the output is 0.6144 for 2 clusters, indicating that this configuration likely provides the most effective separation of data points among the evaluated options.

For two clusters, the silhouette score and Davies-Bouldin index both suggest relatively good clustering performance. But as the number of clusters increases to three and four, both metrics indicate a decline in clustering quality, with lower silhouette scores and higher Davies-Bouldin indices, pointing to less distinct and less cohesive clusters.

Implementation and evaluation of clustering

Since two clusters seem to be the best choice for this dataset and set of features, let's run the model with n_clusters=2:

```python
# Perform K-means clustering with a chosen number of clusters
kmeans = KMeans(n_clusters=2, n_init=10, random_state=0)  # Explicitly setting
n_init to 10
data2['class'] = kmeans.fit_predict(features_scaled)

# Display the first few rows of the dataframe to verify the 'class' column
data2
```

Once again, as shown in the *2. Exploratory data analysis* section, the correlation between complaints and exiting is established, as shown in the excerpt of the pandas DataFrame in *Figure 6.5*:

Exited	Complain	Satisfaction Score	Card Type	Point Earned	class
1	1	2	DIAMOND	464	0
0	1	3	DIAMOND	456	0
1	1	3	DIAMOND	377	0
0	0	5	GOLD	350	1
0	0	5	GOLD	425	1
...	...	...	...	...	...
0	0	1	DIAMOND	300	1
0	0	5	PLATINUM	771	1
1	1	3	SILVER	564	0
1	1	2	GOLD	339	0
0	0	3	DIAMOND	911	1

Figure 6.5: Excerpt of the output of k-means clustering

The first cluster is `class=0`, which represents customers who complained (`Complain`) and left (`Exited`) the bank.

If we count the rows for which `Sum where 'class' == 0` and `'Exited' == 1`, we will obtain a strong correlation between complaints and customers leaving the bank:

```python
# 1. Sum where 'class' == 0
sum_class_0 = (data2['class'] == 0).sum()
# 2. Sum where 'class' == 0 and 'Complain' == 1
sum_class_0_complain_1 = data2[(data2['class'] == 0) & (data2['Complain'] ==
1)].shape[0]
# 3. Sum where 'class' == 0 and 'Exited' == 1
sum_class_0_exited_1 = data2[(data2['class'] == 0) & (data2['Exited'] == 1)].
shape[0]


# Print the results
print(f"Sum of 'class' == 0: {sum_class_0}")
print(f"Sum of 'class' == 0 and 'Complain' == 1: {sum_class_0_complain_1}")
print(f"Sum of 'class' == 0 and 'Exited' == 1: {sum_class_0_exited_1}")
```

The output confirms that complaints and churn (customers leaving the bank) are closely related:

```
Sum of 'class' == 0: 2039
Sum of 'class' == 0 and 'Complain' == 1: 2036
Sum of 'class' == 0 and 'Exited' == 1: 2037
```

The following cell for the second class where `'class' == 1 and 'Complain' == 1` confirms that few customers that complain stay with the bank:

```
# 2. Sum where 'class' == 1 and 'Complain' == 1
sum_class_1_complain_1 = data2[(data2['class'] == 1) & (data2['Complain'] ==
1)].shape[0]
```

The output is consistent with the correlations we have observed:

```
Sum of 'class' == 1: 7961
Sum of 'class' == 1 and 'Complain' == 1: 8
Sum of 'class' == 1 and 'Exited' == 1: 1
```

We saw that finding the features that could help us keep customers is challenging with classical methods that can be effective. However, our strategy will now be to transform the customer records into vectors with OpenAI and query a Pinecone index to find deeper patterns within the dataset with queries that don't exactly match the dataset.

Pipeline 2: Scaling a Pinecone index (vector store)

The goal of this section is to build a Pinecone index with our dataset and scale it from 10,000 records up to 1,000,000 records. Although we are building on the knowledge acquired in the previous chapters, the essence of scaling is different from managing sample datasets.

The clarity of each process of this pipeline is deceptively simple: data preparation, embedding, uploading to a vector store, and querying to retrieve documents. We have already gone through each of these processes in *Chapters 2* and *3*.

Furthermore, beyond implementing Pinecone instead of Deep Lake and using OpenAI models in a slightly different way, we are performing the same functions as in *Chapters 2*, *3*, and *4* for the vector store phase:

1. **Data preparation:** We will start by preparing our dataset using Python for chunking.
2. **Chunking and embedding:** We will chunk the prepared data and then embed the chunked data.
3. **Creating the Pinecone index:** We will create a Pinecone index (vector store).
4. **Upserting:** We will upload the embedded documents (in this case, customer records) and the text of each record as metadata.
5. **Querying the Pinecone index:** Finally, we will run a query to retrieve relevant documents to prepare *Pipeline 3: RAG generative AI*.

Take all the time you need, if necessary, to go through *Chapters 2,3*, and *4* again for the data preparation, chunking, embedding, and querying functions.

We know how to implement each phase because we've already done that with Deep Lake, and Pinecone is a type of vector store, too. So, what's the issue here? The real issue is the hidden real-life project challenges on which we will focus, starting with the size, cost, and operations involved.

The challenges of vector store management

Usually, we begin a section by jumping into the code. That's fine for small volumes, but scaling requires project management decisions before getting started! Why? When we run a program with a bad decision or an error on small datasets, the consequences are limited. But scaling is a different story! The fundamental principle and risk of scaling is that errors are scaled exponentially, too.

Let's list the pain points you must face before running a single line of code. You can apply this methodology to any platform or model. However, we have limited the platforms in this chapter to OpenAI and Pinecone to focus on processes, not platform management. Using other platforms involves careful risk management, which isn't the objective of this chapter.

Let's begin with OpenAI models:

- **OpenAI models for embedding**: OpenAI continually improves and offers new models for embedding. Make sure you examine the characteristics of each one before embedding, including speed, cost, input limits, and API call rates, at `https://platform.openai.com/docs/models/embeddings`.

- **OpenAI models for generation**: OpenAI continually releases new models and abandons older ones. Google does the same. Think of these models as racing cars. Can you win a race today with a 1930 racing car? When scaling, you need the most efficient models. Check the speed, cost, input limits, output size, and API call rates at `https://platform.openai.com/docs/models`.

This means that you must continually take the evolution of models into account for speed and cost reasons when scaling. Then, beyond technical considerations, you must have a real-time view of the pay-as-you-go billing perspective and technical constraints, such as:

- **Billing management**: `https://platform.openai.com/settings/organization/billing/overview`

- **Limits including rate limits**: `https://platform.openai.com/settings/organization/limits`

Now, let's examine Pinecone constraints once you have created an account:

- **Cloud and region:** The choice of the cloud (AWS, Google, or other) and region (location of the serverless storage) have pricing implications.

- **Usage:** This includes read units, write units, and storage costs, including cloud backups. Read more at `https://docs.pinecone.io/guides/indexes/back-up-an-index`.

You must continually monitor the price and usage of Pinecone as for any other cloud environment. You can do so using these links: https://www.pinecone.io/pricing/ and https://docs.pinecone.io/guides/operations/monitoring.

The scenario we are implementing is one of many other ways of achieving the goals in this chapter with other platforms and frameworks. However, the constraints are invariants, including pricing, usage, speed performances, and limits.

Let's now implement *Pipeline 2* by focusing on the pain points beyond the functionality we have already explored in previous chapters. You may open Pipeline_2_Scaling_a_Pinecone_Index.ipynb in the GitHub repository. The program begins with installing the environment.

Installing the environment

As mentioned earlier, the program is limited to Pinecone and OpenAI, which has the advantage of avoiding any intermediate software, platforms, and constraints. Store your API keys in a safe location. In this case, the API keys are stored on Google Drive:

```
#API Key
#Store your key in a file and read it(you can type it directly in the notebook
but it will be visible for somebody next to you)
from google.colab import drive
drive.mount('/content/drive')
```

Now, we install OpenAI and Pinecone:

```
!pip install openai==1.40.3
!pip install pinecone-client==5.0.1
```

Finally, the program initializes the API keys:

```
f = open("drive/MyDrive/files/pinecone.txt", "r")
PINECONE_API_KEY=f.readline()
f.close()
f = open("drive/MyDrive/files/api_key.txt", "r")
API_KEY=f.readline()
f.close()

#The OpenAI Key
import os
import openai
os.environ['OPENAI_API_KEY'] =API_KEY
openai.api_key = os.getenv("OPENAI_API_KEY")
```

The program now processes the Bank Customer Churn dataset.

Processing the dataset

This section will focus on preparing the dataset for chunking, which splits it into optimized chunks of text to embed. The program first retrieves the data1.csv dataset that we prepared and saved in the *Pipeline 1: Collecting and preparing the dataset* section of this chapter:

```
!cp /content/drive/MyDrive/files/rag_c6/data1.csv /content/data1.csv
```

Then, we load the dataset in a pandas DataFrame:

```
import pandas as pd
# Load the CSV file
file_path = '/content/data1.csv'
data1 = pd.read_csv(file_path)
```

We make sure that the 10,000 lines of the dataset are loaded:

```
# Count the chunks
number_of_lines = len(data1)
print("Number of lines: ",number_of_lines)
```

The output confirms that the lines are indeed present:

```
Number of lines:  10000
```

The following code is important in this scenario. Each line that represents a customer record will become a line in the output_lines list:

```
import pandas as pd

# Initialize an empty list to store the lines
output_lines = []

# Iterate over each row in the DataFrame
for index, row in data1.iterrows():
    # Create a list of "column_name: value" for each column in the row
    row_data = [f"{col}: {row[col]}" for col in data1.columns]
    # Join the list into a single string separated by spaces
    line = ' '.join(row_data)
    # Append the line to the output list
    output_lines.append(line)

# Display or further process `output_lines` as needed
for line in output_lines[:5]:  # Displaying first 5 lines for preview
    print(line)
```

The output shows that each line in the output_lines list is a separate customer record text:

```
CustomerId: 15634602 CreditScore: 619 Age: 42 Tenure: 2 Balance: 0.0
NumOfProducts: 1 HasCrCard: 1 IsActiveMember: 1 EstimatedSalary: 101348.88
Exited: 1 Complain: 1 Satisfaction Score: 2 Card Type: DIAMOND Point Earned:
464...
```

We are sure that each line is a separate pre-chunk with a clearly defined customer record. Let's now copy output_lines to lines for the chunking process:

```
lines = output_lines.copy()
```

The program runs a quality control on the lines list to make sure we haven't lost a line in the process:

```
# Count the lines
number_of_lines = len(lines)
print("Number of lines: ",number_of_lines)
```

The output confirms that 10,000 lines are present:

```
Number of lines:  10000
```

And just like that, the data is ready to be chunked.

Chunking and embedding the dataset

In this section, we will chunk and embed the pre-chunks in the lines list. Building a pre-chunks list with structured data is not possible every time, but when it is, it increases a model's traceability, clarity, and querying performance. The chunking process is straightforward.

Chunking

The practice of chunking pre-chunks is important for dataset management. We can create our chunks from a list of pre-chunks stored as lines:

```
# Initialize an empty list for the chunks
chunks = []

# Add each line as a separate chunk to the chunks list
for line in lines:
    chunks.append(line)  # Each line becomes its own chunk

# Now, each line is treated as a separate chunk
print(f"Total number of chunks: {len(chunks)}")
```

The output shows that we have not lost any data during the process:

```
Total number of chunks: 10000
```

So why bother creating chunks and not just use the lines directly? In many cases, lines may require additional quality control and processing, such as data errors that somehow slipped through in the previous steps. We might even have a few chunks that exceed the input limit (which is continually evolving) of an embedding model at a given time.

To better understand the structure of the chunked data, you can examine the length and content of the chunks using the following code:

```
# Print the length and content of the first 10 chunks
for i in range(3):
    print(len(chunks[i]))
    print(chunks[i])
```

The output will help a human controller visualize the chunked data, providing a snapshot like so:

```
224
CustomerId: 15634602 CreditScore: 619 Age: 42 Tenure: 2 Balance: 0.0
NumOfProducts: 1 HasCrCard: 1 IsActiveMember: 1 EstimatedSalary: 101348.88
Exited: 1 Complain: 1 Satisfaction Score: 2 Card Type: DIAMOND Point Earned:
464...
```

The chunks will now be embedded.

Embedding

This section will require careful testing and consideration of the issues. We will realize that *scaling requires more thinking than doing*. Each project will require specific amounts of data through design and testing to provide effective responses. We must also take into account the cost and benefit of each component of the pipeline. For example, initializing the embedding model is no easy task!

At the time of writing, OpenAI offers three embedding models that we can test:

```
import openai
import time

embedding_model="text-embedding-3-small"
#embedding_model="text-embedding-3-large"
#embedding_model="text-embedding-ada-002"
```

In this section, we will use text-embedding-3-small. However, you can evaluate the other models by uncommenting the code. The embedding function will accept the model you select:

```
# Initialize the OpenAI client
client = openai.OpenAI()

def get_embedding(text, model=embedding_model):
    text = text.replace("\n", " ")
```

```
    response = client.embeddings.create(input=[text], model=model)
    embedding = response.data[0].embedding
    return embedding
```

Make sure to check the cost and features of each embedding model before running one of your choice: https://platform.openai.com/docs/guides/embeddings/embedding-models.

The program now embeds the chunks, but the embedding process requires strategic choices, particularly to manage large datasets and API rate limits effectively. In this case, we will create batches of chunks to embed:

```
import openai
import time

# Initialize the OpenAI client
client = openai.OpenAI()

# Initialize variables
start_time = time.time()  # Start timing before the request
chunk_start = 0
chunk_end = 1000
pause_time = 3
embeddings = []
counter = 1
```

We will embed 1,000 chunks at a time with `chunk_start = 0` and `chunk_end = 1000`. To avoid possible OpenAI API rate limits, `pause_time = 3` was added to pause for 3 seconds between each batch. We will store the embeddings in `embeddings = []` and count the batches starting with `counter = 1`.

The code is divided into three main parts, as explained in the following excerpts:

- Iterating through all the chunks with batches:

  ```
  while chunk_end <= len(chunks):
      # Select the current batch of chunks
      chunks_to_embed = chunks[chunk_start:chunk_end]…
  ```

- Embedding a batch of `chunks_to_embed`:

  ```
  for chunk in chunks_to_embed:
      embedding = get_embedding(chunk, model=embedding_model)
      current_embeddings.append(embedding)…
  ```

- Updating the start and end values of the chunks to embed for the next batch:

  ```
  # Update the chunk indices
  chunk_start += 1000
  chunk_end += 1000
  ```

A function was added in case the batches are not perfect multiples of the batch size:

```
# Process the remaining chunks if any
if chunk_end < len(chunks):
    remaining_chunks = chunks[chunk_end:]
    remaining_embeddings = [get_embedding(chunk, model=embedding_model) for
chunk in remaining_chunks]
    embeddings.extend(remaining_embeddings)
```

The output displays the counter and the processing time:

```
All chunks processed.
Batch 1 embedded.
...
Batch 10 embedded.
Response Time: 2689.46  seconds
```

The response time may seem long and may vary for each run, but that is what scaling is all about! We cannot expect to process large volumes of data in a very short time and not face performance challenges.

We can display an embedding if we wish to check that everything went well:

```
print("First embedding:", embeddings[0])
```

The output displays the embedding:

```
First embedding: [-0.024449337273836136, -0.00936567410826683,…
```

Let's verify if we have the same number of text chunks (customer records) and vectors (embeddings):

```
# Check the lengths of the chunks and embeddings
num_chunks = len(chunks)
print(f"Number of chunks: {num_chunks}")
print(f"Number of embeddings: {len(embeddings)}")
```

The output confirms that we are ready to move to Pinecone:

```
Number of chunks: 10000
Number of embeddings: 10000
```

We have now chunked and embedded the data. We will duplicate the data to simulate scaling in this notebook.

Duplicating data

We will duplicate the chunked and embedded data; this way, you can simulate volumes without paying for the OpenAI embeddings. The cost of the embedding data and the time performances are linear. So we can simulate scaling with a corpus of 50,000 data points, for example, and extrapolate the response times and cost to any size we need.

The code is straightforward. We first determine the number of times we want to duplicate the data:

```
# Define the duplication size
dsize = 5  # You can set this to any value between 1 and n as per your
experimentation requirements
total=dsize * len(chunks)
print("Total size", total)
```

The program will then duplicate the chunks and the embeddings:

```
# Initialize new lists for duplicated chunks and embeddings
duplicated_chunks = []
duplicated_embeddings = []


# Loop through the original lists and duplicate each entry
for i in range(len(chunks)):
    for _ in range(dsize):
        duplicated_chunks.append(chunks[i])
        duplicated_embeddings.append(embeddings[i])
```

The code then checks if the number of chunks fits the number of embeddings:

```
# Checking the lengths of the duplicated lists
print(f"Number of duplicated chunks: {len(duplicated_chunks)}")
print(f"Number of duplicated embeddings: {len(duplicated_embeddings)}")
```

Finally, the output confirms that we duplicated the data five times:

```
Total size 50000
Number of duplicated chunks: 50000
Number of duplicated embeddings: 50000
```

50,000 data points is a good volume to begin with, giving us the necessary data to populate a vector store. Let's now create the Pinecone index.

Creating the Pinecone index

The first step is to make sure our API key is initialized with the name of the variable we prefer and then create a Pinecone instance:

```
import os
from pinecone import Pinecone, ServerlessSpec
# initialize connection to pinecone (get API key at app.pinecone.io)
api_key = os.environ.get('PINECONE_API_KEY') or 'PINECONE_API_KEY'
pc = Pinecone(api_key=PINECONE_API_KEY)
```

The Pinecone instance, pc, has been created. Now, we will choose the index name, our cloud, and region:

```
from pinecone import ServerlessSpec
index_name = [YOUR INDEX NAME] #'bank-index-900'for example
cloud = os.environ.get('PINECONE_CLOUD') or 'aws'
region = os.environ.get('PINECONE_REGION') or 'us-east-1'
spec = ServerlessSpec(cloud=cloud, region=region)
```

We have now indicated that we want a serverless cloud instance (spec) with AWS in the 'us-east-1' location. We are ready to create the index (the type of vector store) named 'bank-index-50000' with the following code:

```
import time
import pinecone
# check if index already exists (it shouldn't if this is first time)
if index_name not in pc.list_indexes().names():
    # if does not exist, create index
    pc.create_index(
        index_name,
        dimension=1536,   #Dimension of the embedding model
        metric='cosine',
        spec=spec
    )
    # wait for index to be initialized
    time.sleep(1)
# connect to index
index = pc.Index(index_name)
# view index stats
index.describe_index_stats()
```

We added the following two parameters to index_name and spec:

- dimension=1536 represents the length of the embeddings vector that you can adapt to the embedding model of your choice.
- metric='cosine' is the metric we will use for vector similarity between the embedded vectors. You can also choose other metrics, such as Euclidean distance: https://www.pinecone.io/learn/vector-similarity/.

When the index is created, the program displays the description of the index:

```
{'dimension': 1536,
 'index_fullness': 0.0,
 'namespaces': {},
 'total_vector_count': 0}
```

The vector count and index fullness are 0 since we haven't been populating the vector store. Great, now we are ready to upsert!

Upserting

The section's goal is to populate the vector store with our 50,000 embedded vectors and their associated metadata (chunks). The objective is to fully understand the scaling process and use synthetic data to reach the 50,000+ vector level. You can go back to the previous section and duplicate the data up to any value you wish. However, bear in mind that the upserting time to a Pinecone index is linear. You simply need to extrapolate the performances to the size you want to evaluate to obtain the approximate time it would take. Check the Pinecone pricing before running the upserting process: https://www.pinecone.io/pricing/.

We will populate (upsert) the vector store with three fields:

- ids: Contains a unique identifier for each chunk, which will be a counter we increment as we upsert the data
- embedding: Contains the vectors (embedded chunks) we created
- chunks: Contains the chunks in plain text, which is the metadata

The code will populate the data in batches. Let's first define the batch upserting function:

```
# upsert function
def upsert_to_pinecone(data, batch_size):
    for i in range(0, len(data), batch_size):
        batch = data[i:i+batch_size]
        index.upsert(vectors=batch)
        #time.sleep(1)  # Optional: add delay to avoid rate limits
```

We will measure the time it takes to process our corpus:

```
import pinecone
import time
import sys
start_time = time.time()  # Start timing before the request
```

Now, we create a function that will calculate the size of the batches and limit them to 4 MB, which is close to the present Pinecone upsert batch size limit:

```
# Function to calculate the size of a batch
def get_batch_size(data, limit=4000000):  # limit set slightly below 4MB to be
safe
    total_size = 0
    batch_size = 0
    for item in data:
        item_size = sum([sys.getsizeof(v) for v in item.values()])
        if total_size + item_size > limit:
            break
        total_size += item_size
```

```
            batch_size += 1
        return batch_size
```

We can now create our upsert function:

```
def batch_upsert(data):
    total = len(data)
    i = 0
    while i < total:
        batch_size = get_batch_size(data[i:])
        batch = data[i:i + batch_size]
        if batch:
            upsert_to_pinecone(batch,batch_size)
            i += batch_size
            print(f"Upserted {i}/{total} items...")  # Display current progress
        else:
            break
    print("Upsert complete.")
```

We need to generate unique IDs for the data we upsert:

```
# Generate IDs for each data item
ids = [str(i) for i in range(1, len(duplicated_chunks) + 1)]
```

We will create the metadata to upsert the dataset to Pinecone:

```
# Prepare data for upsert
data_for_upsert = [
    {"id": str(id), "values": emb, "metadata": {"text": chunk}}
    for id, (chunk, emb) in zip(ids, zip(duplicated_chunks, duplicated_
embeddings))
]
```

We now have everything we need to upsert in `data_for_upsert`:

- `"id": str(ids[i])` contains the IDs we created with the seed.
- `"values": emb` contains the chunks we embedded into vectors.
- `"metadata": {"text": chunk}` contains the chunks we embedded.

We now run the batch upsert process:

```
# Upsert data in batches
batch_upsert(data_for_upsert)
```

Finally, we measure the response time:

```
response_time = time.time() - start_time  # Measure response time
```

```
print(f"Upsertion response time: {response_time:.2f} seconds")  # Print
response time
```

The output contains useful information that shows the batch progression:

```
Upserted 316/50000 items...
Upserted 632/50000 items...
Upserted 948/50000 items...

...

Upserted 49612/50000 items...
Upserted 49928/50000 items...
Upserted 50000/50000 items...
Upsert complete.
Upsertion response time: 560.66 seconds
```

The time shows that it takes just under one minute (56 seconds) per 10,000 data points. You can try a larger corpus. The time should remain linear.

We can also view the Pinecone index statistics to see how many vectors were uploaded:

```
print("Index stats")
print(index.describe_index_stats(include_metadata=True))
```

The output confirms that the upserting process was successful:

```
Index stats
{'dimension': 1536,
 'index_fullness': 0.0,
 'namespaces': {'': {'vector_count': 50000}},
 'total_vector_count': 50000}
```

The upsert output shows that we upserted 50,000 data points but the output shows less, most probably due to duplicates within the data.

Querying the Pinecone index

The task now is to verify the response times with a large Pinecone index. Let's create a function to query the vector store and display the results:

```
# Print the query results along with metadata
def display_results(query_results):
  for match in query_results['matches']:
    print(f"ID: {match['id']}, Score: {match['score']}")
    if 'metadata' in match and 'text' in match['metadata']:
        print(f"Text: {match['metadata']['text']}")
    else:
        print("No metadata available.")
```

We need an embedding function for the query using the same embedding model as we implemented to embed the chunks of the dataset:

```python
embedding_model = "text-embedding-3-small"
def get_embedding(text, model=embedding_model):
    text = text.replace("\n", " ")
    response = client.embeddings.create(input=[text], model=model)
    embedding = response.data[0].embedding
    return embedding
```

We can now query the Pinecone vector store to conduct a unit test and display the results and response time. We first initialize the OpenAI client and start time:

```python
import openai
# Initialize the OpenAI client
client = openai.OpenAI()
print("Querying vector store")
start_time = time.time()  # Start timing before the request
```

We then query the vector store with a customer profile that does not exist in the dataset:

```python
query_text = "Customer Robertson CreditScore 632Age 21 Tenure 2Balance
0.0NumOfProducts 1HasCrCard 1IsActiveMember 1EstimatedSalary 99000 Exited
1Complain 1Satisfaction Score 2Card Type DIAMONDPoint Earned 399"
```

The query is embedded with the same model as the one used to embed the dataset:

```python
query_embedding = get_embedding(query_text,model=embedding_model)
```

We run the query and display the output:

```python
query_results = index.query(vector=query_embedding, top_k=1, include_
metadata=True)  # Request metadata
#print("raw query_results",query_results)
print("processed query results")
display_results(query_results) #display results
response_time = time.time() - start_time                    # Measure response time
print(f"Querying response time: {response_time:.2f} seconds")  # Print response
time
```

The output displays the query response and time:

```
Querying vector store
Querying vector store
processed query results
ID: 46366, Score: 0.823366046
```

```
Text: CustomerId: 15740160 CreditScore: 616 Age: 31 Tenure: 1 Balance: 0.0
NumOfProducts: 2 HasCrCard: 1 IsActiveMember: 1 EstimatedSalary: 54706.75
Exited: 0 Complain: 0 Satisfaction Score: 3 Card Type: DIAMOND Point Earned:
852
Querying response time: 0.74 seconds
```

We can see that the response quality is satisfactory because it found a similar profile. The time is excellent: 0.74 seconds. When reaching a 1,000,000 vector count, for example, the response time should still be constant at less than a second. That is the magic of the Pinecone index!

If we go to our organization on Pinecone, https://app.pinecone.io/organizations/, and click on our index, we can monitor our statistics, analyze our usage, and more, as illustrated here:

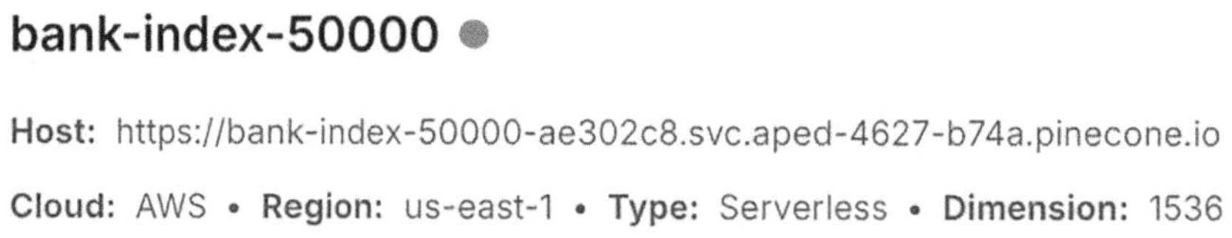

bank-index-50000 ●

Host: https://bank-index-50000-ae302c8.svc.aped-4627-b74a.pinecone.io

Cloud: AWS • **Region:** us-east-1 • **Type:** Serverless • **Dimension:** 1536

Figure 6.6: Visualizing the Pinecone index vector count in the Pinecone console

Our Pinecone index is now ready to augment inputs and generate content.

Pipeline 3: RAG generative AI

In this section, we will use RAG generative AI to automate a customized and engaging marketing message to the customers of the bank to encourage them to remain loyal. We will be building on our programs on data preparation and Pinecone indexing; we will leverage the Pinecone vector database for advanced search functionalities. We will choose a target vector that represents a market segment to query the Pinecone index. The response will be processed to extract the top k similar vectors. We will then augment the user input with this target market to ask OpenAI to make recommendations to the market segment targeted with customized messages.

You may open Pipeline-3_RAG_Generative AI.ipynb on GitHub. The first code section in this notebook, *Installing the environment*, is the same as in 2-Pincone_vector_store-1M.ipynb, built in the *Pipeline 2: Scaling a Pinecone index (vector store)* section earlier in this chapter. The *Pinecone index* in the second code section is also the same as in 2-Pincone_vector_store-1M.ipynb. However, this time, the Pinecone index code checks whether a Pinecone index exists and connects to it if it does, rather than creating a new index.

Let's run an example of RAG with GPT-4o.

RAG with GPT-4o

In this section of the code, we will query the Pinecone vector store, augment the user input, and generate a response with GPT-4o. It is the same process as with Deep Lake and an OpenAI generative model in *Chapter 3, Building Index-Based RAG with LlamaIndex, Deep Lake, and OpenAI*, for example. However, the nature and usage of the Pinecone query is quite different in this case for the following reasons:

- **Target vector:** The user input is not a question in the classical sense. In this case, it is a target vector representing the profile of a market segment.

- **Usage:** The usage isn't to augment the generative AI in the classical dialog sense (questions, summaries). In this case, we expect GPT-4o to write an engaging, customized email to offer products and services.

- **Query time:** Speed is critical when scaling an application. We will measure the query time on the Pinecone index that contains 1,000,000+ vectors.

Querying the dataset

We will need an embedding function to embed the input. We will simplify and use the same embedding model we used in the *Embedding* section of *Pipeline 2: Scaling a Pinecone index (vector store)* for compatibility reasons:

```
import openai
import time


embedding_model= "text-embedding-3-small"
# Initialize the OpenAI client
client = openai.OpenAI()
def get_embedding(text, model=embedding_model):
    text = text.replace("\n", " ")
    response = client.embeddings.create(input=[text], model=model)
    embedding = response.data[0].embedding
    return embedding
```

We are now ready to query the Pinecone index.

Querying a target vector

A target vector represents a market segment that a marketing team wants to focus on for recommendations to increase customer loyalty. Your imagination and creativity are the only limits! Usually, the marketing team will be part of the design team for this pipeline. You might want to organize workshops to try various scenarios until the marketing team is satisfied. If you are part of the marketing team, then you want to help design target vectors. In any case, human insights into our adaptive creativity will lead to many ways of organizing target vectors and queries.

In this case, we will target a market segment of customers around the age of 42 (Age 42). We don't need the age to be strictly 42 or an age bracket. We'll let AI do the work for us. We are also targeting a customer that has a 100,000+ (EstimatedSalary 101348.88) estimated salary, which would be a loss for the bank. We're choosing a customer who has complained (Complain 1) and seems to be exiting (Exited 1) the bank. Let's suppose that Exited 1, in this scenario, means that the customer has made a request to close an account but it hasn't been finalized yet. Let's also consider that the marketing department chose the target vector.

query_text represents the customer profiles we are searching for:

```
import time
start_time = time.time()  # Start timing before the request
# Target vector
    "

# Target vector
query_text = "Customer Henderson CreditScore 599 Age 37Tenure 2Balance
0.0NumOfProducts 1HasCrCard 1IsActiveMember 1EstimatedSalary 107000.88Exited
1Complain 1Satisfaction Score 2Card Type DIAMONDPoint Earned 501"

query_embedding = get_embedding(text,model=embedding_model)
```

We have embedded the query. Let's now retrieve the top-k customer profiles that fit the target vector and parse the result:

```
# Perform the query using the embedding
query_results = index.query(
    vector=query_embedding,
    top_k=5,
    include_metadata=True,
)
```

We now print the response and the metadata:

```
# Print the query results along with metadata
print("Query Results:")
for match in query_results['matches']:
    print(f"ID: {match['id']}, Score: {match['score']}")
    if 'metadata' in match and 'text' in match['metadata']:
        print(f"Text: {match['metadata']['text']}")
    else:
        print("No metadata available.")

response_time = time.time() - start_time              # Measure response time
print(f"Querying response time: {response_time:.2f} seconds")  # Print response
time
```

The result is parsed to find the top-k matches to display their scores and content, as shown in the following output:

```
Query Results:
ID: 46366, Score: 0.854999781
Text: CustomerId: 15740160 CreditScore: 616 Age: 31 Tenure: 1 Balance: 0.0
NumOfProducts: 2 HasCrCard: 1 IsActiveMember: 1 EstimatedSalary: 54706.75
Exited: 0 Complain: 0 Satisfaction Score: 3 Card Type: DIAMOND Point Earned:
852
Querying response time: 0.63 seconds
```

We have retrieved valuable information:

- **Ranking** through the top-k vectors that match the target vector. From one to another, depending on the target vector, the ranking will be automatically recalculated by the OpenAI generative AI model.

- **Score metric** through the score provided. A score is returned providing a metric for the response.

- **Content** that contains the top-ranked and best scores.

It's an all-in-one automated process! AI is taking us to new heights but we, of course, need human control to confirm the output, as described in the previous chapter on human feedback.

We now need to extract the relevant information to augment the input.

Extracting relevant texts

The following code goes through the top-ranking vectors, searches for the matching text metadata, and combines the content to prepare the augmentation phase:

```
relevant_texts = [match['metadata']['text'] for match in query_
results['matches'] if 'metadata' in match and 'text' in match['metadata']]

# Join all items in the list into a single string separated by a specific
delimiter (e.g., a newline or space)
combined_text = '\n'.join(relevant_texts)  # Using newline as a separator for
readability
print(combined_text)
```

The output displays combined_text, relevant text we need to augment the input:

```
CustomerId: 15740160 CreditScore: 616 Age: 31 Tenure: 1 Balance: 0.0
NumOfProducts: 2 HasCrCard: 1 IsActiveMember: 1 EstimatedSalary: 54706.75
Exited: 0 Complain: 0 Satisfaction Score: 3 Card Type: DIAMOND Point Earned:
852
```

We are now ready to augment the prompt before AI generation.

Augmented prompt

We will now engineer our prompt by adding three texts:

- query_prompt: The instructions for the generative AI model
- query_text: The target vector containing the target profile chosen by the marketing team
- combined_context: The concentrated metadata text of the similar vectors selected by the query

itext contains these three variables:

```
# Combine texts into a single string, separated by new lines
combined_context = "\n".join(relevant_texts)
#prompt
query_prompt="I have this customer bank record with interesting information
on age, credit score and more and similar customers. What could I suggest to
keep them in my bank in an email with an url to get new advantages based on the
fields for each Customer ID:"
itext=query_prompt+ query_text+combined_context
# Augmented input
print("Prompt for the Generative AI model:", itext)
```

The output is the core input for the generative AI model:

```
Prompt for GPT-4: I have this customer bank record with interesting information
on age, credit score and more and similar customers. What could I suggest to
keep them in my bank in an email with an url to get new advantages based on the
fields for each Customer ID:…
```

We can now prepare the request for the generative AI model.

Augmented generation

In this section, we will submit the augmented input to an OpenAI generative AI model. The goal is to obtain a customized email to send the customers in the Pinecone index marketing segment we obtained through the target vector.

We will first create an OpenAI client and choose GPT-4o as the generative AI model:

```
from openai import OpenAI
client = OpenAI()
gpt_model = "gpt-4o
```

We then introduce a time performance measurement:

```
import time
start_time = time.time()  # Start timing before the request
```

The response time should be relatively constant since we are only sending one request at a time in this scenario. We now begin to create our completion request:

```
response = client.chat.completions.create(
  model=gpt_model,
  messages=[
```

The system role provides general instructions to the model:

```
    {
       "role": "system",
       "content": "You are the community manager can write engaging email based
  on the text you have. Do not use a surname but simply Dear Valued Customer
  instead."
    },
```

The user role contains the engineered itext prompt we designed:

```
    {
       "role": "user",
       "content": itext
    }
  ],
```

Now, we set the parameters for the request:

```
    temperature=0,
    max_tokens=300,
    top_p=1,
    frequency_penalty=0,
    presence_penalty=0
)
```

The parameters are designed to obtain a low random yet "creative" output:

- `temperature=0`: Low randomness in response
- `max_tokens=300`: Limits response length to 300 tokens
- `top_p=1`: Considers all possible tokens; full diversity
- `frequency_penalty=0`: No penalty for frequent word repetition to allow the response to remain open
- `presence_penalty=0`: No penalty for introducing new topics to allow the response to find ideas for our prompt

We send the request and display the response:

```
print(response.choices[0].message.content)
```

The output is satisfactory for this market segment:

```
Subject: Exclusive Benefits Await You at Our Bank!


Dear Valued Customer,


We hope this email finds you well. At our bank, we are constantly striving
to enhance your banking experience and provide you with the best possible
services. We have noticed that you are a valued customer with a DIAMOND card,
and we would like to offer you some exclusive benefits tailored just for you!

Based on your profile, we have identified several opportunities that could
enhance your banking experience:


1. **Personalized Financial Advice**: Our financial advisors are available to
help you make the most of your finances. Whether it's planning for the future
or managing your current assets, we are here to assist you.


2. **Exclusive Rewards and Offers**: As a DIAMOND cardholder, you are eligible
for special rewards and offers. Earn more points and enjoy exclusive discounts
on various products and services.


3. **Enhanced Credit Options**: With your current credit score, you may qualify
for better credit options. We can help you explore these opportunities to
improve your financial standing.


4. **Complimentary Financial Health Check**: We understand the importance of
financial well-being. Schedule a complimentary financial health check to ensure
you are on the right track.


5. **Loyalty Programs**: Participate in our loyalty programs and earn more
points for every transaction. Redeem these points for exciting rewards and
benefits.


To explore these new advantages and more, please visit the following link:
[Exclusive Benefits](https://www.yourbank
```

Since the goal of the marketing team is to convince customers not to leave and to increase their loyalty to the bank, I'd say the email we received as output is good enough. Let's display the time it took to obtain a response:

```
response_time = time.time() - start_time           # Measure response time
print(f"Querying response time: {response_time:.2f} seconds")  # Print response
time
```

The response time is displayed:

```
Querying response time: 2.83 seconds
```

We have successfully produced a customized response based on a target vector. This approach might be sufficient for some projects, whatever the domain. Let's summarize the RAG-driven generative recommendation system built in this chapter and continue our journey.

Summary

This chapter aimed to develop a scaled RAG-driven generative AI recommendation system using a Pinecone index and OpenAI models tailored to mitigate bank customer churn. Using a Kaggle dataset, we demonstrated the process of identifying and addressing factors leading to customer dissatisfaction and account closures. Our approach involved three key pipelines.

When building *Pipeline 1*, we streamlined the dataset by removing non-essential columns, reducing both data complexity and storage costs. Through EDA, we discovered a strong correlation between customer complaints and account closures, which a k-means clustering model further validated. We then designed *Pipeline 2* to prepare our RAG-driven system to generate personalized recommendations. We processed data chunks with an OpenAI model, embedding these into a Pinecone index. Pinecone's consistent upsert capabilities ensured efficient data handling, regardless of volume. Finally, we built *Pipeline 3* to leverage over 1,000,000 vectors within Pinecone to target specific market segments with tailored offers, aiming to boost loyalty and reduce attrition. Using GPT-4o, we augmented our queries to generate compelling recommendations.

The successful application of a targeted vector representing a key market segment illustrated our system's potential to craft impactful customer retention strategies. However, we can improve the recommendations by expanding the Pinecone index into a multimodal knowledge base, which we will implement in the next chapter.

Questions

1. Does using a Kaggle dataset typically involve downloading and processing real-world data for analysis?
2. Is Pinecone capable of efficiently managing large-scale vector storage for AI applications?
3. Can k-means clustering help validate relationships between features such as customer complaints and churn?
4. Does leveraging over a million vectors in a database hinder the ability to personalize customer interactions?
5. Is the primary objective of using generative AI in business applications to automate and improve decision-making processes?
6. Are lightweight development environments advantageous for rapid prototyping and application development?
7. Can Pinecone's architecture automatically scale to accommodate increasing data loads without manual intervention?

8. Is generative AI typically employed to create dynamic content and recommendations based on user data?

9. Does the integration of AI technologies like Pinecone and OpenAI require significant manual configuration and maintenance?

10. Are projects that use vector databases and AI expected to effectively handle complex queries and large datasets?

References

- Pinecone documentation: `https://docs.pinecone.io/guides/get-started/quickstart`
- OpenAI embedding and generative models: `https://platform.openai.com/docs/models`

Further reading

- Han, Y., Liu, C., & Wang, P. (2023). *A comprehensive survey on vector database: Storage and retrieval technique, challenge.*

Unlock this book's exclusive benefits now

This book comes with additional benefits designed to elevate your learning experience.

Note: Have your purchase invoice ready before you begin. `https://www.packtpub.com/unlock/9781836200918`

7

Building Scalable Knowledge-Graph-Based RAG with Wikipedia API and LlamaIndex

Scaled datasets can rapidly become challenging to manage. In real-life projects, data management generates more headaches than AI! Project managers, consultants, and developers constantly struggle to obtain the necessary data to get any project running, let alone a RAG-driven generative AI application. Data is often unstructured before it becomes organized in one way or another through painful decision-making processes. Wikipedia is a good example of how scaling data leads to mostly reliable but sometimes incorrect information. Real-life projects often evolve the way Wikipedia does. Data keeps piling up in a company, challenging database administrators, project managers, and users.

One of the main problems is seeing how large amounts of data fit together, and **knowledge graphs** provide an effective way of visualizing the relationships between different types of data. This chapter begins by defining the architecture of a knowledge base ecosystem designed for RAG-driven generative AI. The ecosystem contains three pipelines: data collection, populating a vector store, and running a knowledge graph index-based RAG program. We will then build *Pipeline 1: Collecting and preparing the documents*, in which we will build an automated Wikipedia retrieval program with the Wikipedia API. We will simply choose a topic based on a Wikipedia page and then let the program retrieve the metadata we need to collect and prepare the data. The system will be flexible and allow you to choose any topic you wish. The use case to first run the program is a marketing knowledge base for students who want to upskill for a new job, for example. The next step is to build *Pipeline 2: Creating and populating the Deep Lake vector store*. We will load the data in a vector store leveraging Deep Lake's in-built automated chunking and OpenAI embedding functionality. We will peek into the dataset to explore how this marvel of technology does the job.

Finally, we will build *Pipeline 3: Knowledge graph index-based RAG*, where LlamaIndex will automatically build a knowledge graph index. It will be exciting to see how the index function churns through our data and produces a graph showing semantic relationships contained in our data. We will then query the graph with LlamaIndex's in-built OpenAI functionality to automatically manage user inputs and produce a response. We will also see how re-ranking can be done and implement metrics to calculate and display the system's performance.

This chapter covers the following topics:

- Defining knowledge graphs
- Implementing the Wikipedia API to prepare summaries and content
- Citing Wikipedia sources in an ethical approach
- Populating a Deep Lake vector store with Wikipedia data
- Building a knowledge graph index with LlamaIndex
- Displaying the LlamaIndex knowledge graph
- Interacting with the knowledge graph
- Generating retrieval responses with the knowledge graph
- Re-ranking the order retrieval responses to choose a better output
- Evaluating and measuring the outputs with metrics

Let's begin by defining the architecture of RAG for knowledge-based semantic search.

The architecture of RAG for knowledge-graph-based semantic search

As established, we will build a graph-based RAG program in this chapter. The graph will enable us to visually map out the relationships between the documents of a RAG dataset. It can be created automatically with LlamaIndex, as we will do in the *Pipeline 3: Knowledge graph index-based RAG* section of this chapter. The program in this chapter will be designed for any Wikipedia topic, as illustrated in the following figure:

RAG for Graph-Based Semantic Search
From a Wikipedia topic to a knowledge graph
interactive LlamaIndex

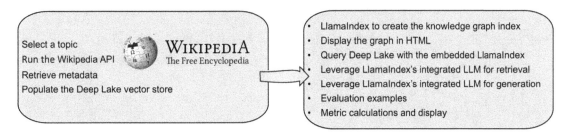

Figure 7.1: From a Wikipedia topic to interacting with a graph-based vector store index

We will first implement a marketing agency for which a knowledge graph can visually map out the complex relationships between different marketing concepts. Then, you can go back and explore any topic you wish once you understand the process. In simpler words, we will implement the three pipelines seamlessly to:

- Select a Wikipedia topic related to *marketing*. Then, you can run the process with the topic of your choice to explore the ecosystem.
- Generate a corpus of Wikipedia pages with the Wikipedia API.
- Retrieve and store the citations for each page.
- Retrieve and store the URLs for each page.
- Retrieve and upsert the content of the URLs in a Deep Lake vector store.
- Build a knowledge base index with LlamaIndex.
- Define a user input prompt.
- Query the knowledge base index.
- Let LlamaIndex's in-built LLM functionality, based on OpenAI's embedding models, produce a response based on the embedded data in the knowledge graph.
- Evaluate the LLM's response with a sentence transformer.
- Evaluate the LLM's response with a human feedback score.
- Provide time metrics for the key functions, which you can extend to other functions if necessary.
- Run metric calculations and display the results.

To attain our goal, we will implement three pipelines leveraging the components we have already built in the previous chapters, as illustrated in the following figure:

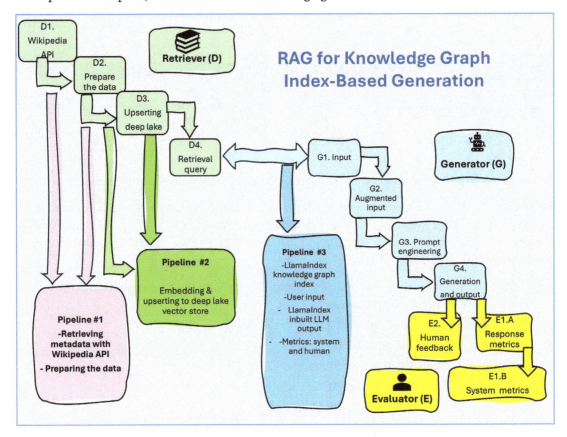

Figure 7.2: Knowledge graph ecosystem for index-based RAG

- **Pipeline 1: Collecting and preparing the documents** will involve building a Wikipedia program using the Wikipedia API to retrieve links from a Wikipedia page and the metadata for all the pages (summary, URL, and citation data). Then, we will load and parse the URLs to prepare the data for upserting.

- **Pipeline 2: Creating and populating the Deep Lake vector store** will embed and upsert parsed content of the Wikipedia pages prepared by *Pipeline 1* to a Deep Lake vector store.

- **Pipeline 3: Knowledge graph index-based RAG** will build the knowledge graph index using embeddings with LlamaIndex and display it. Then, we will build the functionality to query the knowledge base index and let LlamaIndex's in-built LLM generate the response based on the updated dataset.

In this chapter's scenario, we are directly implementing an augmented retrieval system leveraging OpenAI's embedding models more than we are augmenting inputs. This implementation shows the many ways we can improve real-time data retrieval with LLMs. There are no conventional rules. What works, works!

The ecosystem of the three pipelines will be controlled by a scenario that will enable an administrator to either query the vector base or add new Wikipedia pages, as we will implement in this chapter. As such, the architecture of the ecosystem allows for indefinite scaling since it processes and populates the vector dataset one set of Wikipedia pages at a time. The system only uses a CPU and an optimized amount of memory. There are limits to this approach since the LlamaIndex knowledge graph index is loaded with the entire dataset. We can only load portions of the dataset as the vector store grows. Or, we can create one Deep Lake vector store per topic and run queries on multiple datasets. These are decisions to make in real-life projects that require careful decision-making and planning depending on the specific requirements of each project.

We will now dive into the code, beginning a tree-to-graph sandbox.

Building graphs from trees

A graph is a collection of nodes (or vertices) connected by edges (or arcs). Nodes represent entities, and edges represent relationships or connections between these entities. For instance, in our chapter's use case, nodes could represent various marketing strategies, and the edges could show how these strategies are interconnected. This helps new customers understand how different marketing tactics work together to achieve overall business goals, facilitating clearer communication and more effective strategy planning. You can play around with the tree-to-graph sandbox before building the pipelines in this chapter.

You may open `Tree-2-Graph.ipynb` on GitHub. The provided program is designed to visually represent relationships in a tree structure using NetworkX and Matplotlib in Python. It specifically creates a directed graph from given pairs, checks and marks friendships, and then displays this tree with customized visual attributes.

The program first defines the main functions:

- `build_tree_from_pairs(pairs)`: Constructs a directed graph (tree) from a list of node pairs, potentially identifying a root node
- `check_relationships(pairs, friends)`: Checks and prints the friendship status for each pair
- `draw_tree(G, layout_choice, root, friends)`: Visualizes the tree using `matplotlib`, applying different styles to edges based on friendship status and different layout options for node positioning

Then, the program executes the process from tree to graph:

- Node pairs and friendship data are defined.
- The tree is built from the pairs.
- Relationships are checked against the friendship data.
- The tree is drawn using a selected layout, with edges styled differently to denote friendship.

For example, the program first defines a set of node pairs with their pairs of friends:

```
# Pairs
pairs = [('a', 'b'), ('b', 'e'), ('e', 'm'), ('m', 'p'), ('a', 'z'), ('b',
'q')]
friends = {('a', 'b'), ('b', 'e'), ('e', 'm'), ('m', 'p')}
```

Notice that ('a', 'z') are not friends because they are not on the `friends` list. Neither are ('b', 'q'). You can imagine any type of relationship between the pairs, such as the same customer age, similar job, same country, or any other concept you wish to represent. For instance, the `friends` list could contain relationships between friends on social media, friends living in the same country, or anything else you can imagine or need!

The program then builds the tree and checks the relationships:

```
# Build the tree
tree, root = build_tree_from_pairs(pairs)
# Check relationships
check_relationships(pairs, friends)
```

The output shows which pairs are friends and which ones are not:

```
Pair ('a', 'b'): friend
Pair ('b', 'e'): friend
Pair ('e', 'm'): friend
Pair ('m', 'p'): friend
Pair ('a', 'z'): not friend
Pair ('b', 'q'): not friend
```

The output can be used to provide useful information for similarity searches. The program now draws the graph with the 'spring' layout:

```
# Draw the tree
layout_choice = 'spring'  # Define your layout choice here
draw_tree(tree, layout_choice=layout_choice, root=root, friends=friends)
```

The 'spring' layout attracts nodes attracted by edges, simulating the effect of springs. It also ensures that all nodes repel each other to avoid overlapping. You can dig into the draw_tree function to explore and select other layouts listed there. You can also modify the colors and line styles.

In this case, the pairs of friends are represented with solid lines, and the pairs that are not friends are represented with dashes, as shown in the following graph:

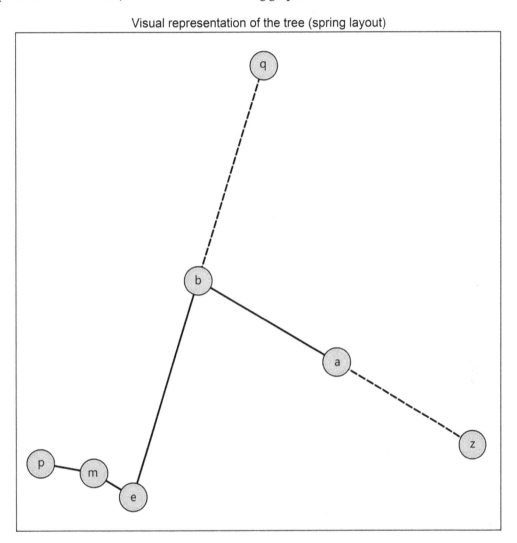

Figure 7.3: Example of a spring layout

You can play with this sandbox graph with different pairs of nodes. If you imagine doing this with hundreds of nodes, you will begin to appreciate the automated functionality we will build in this chapter with LlamaIndex's knowledge graph index!

Let's go from the architecture to the code, starting by collecting and preparing the documents.

Pipeline 1: Collecting and preparing the documents

The code in this section retrieves the metadata we need from Wikipedia, retrieves the documents, cleans them, and aggregates them to be ready for insertion into the Deep Lake vector store. This process is illustrated in the following figure:

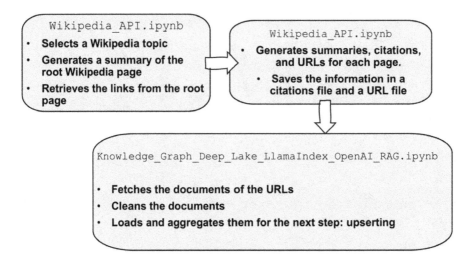

Pipeline 1: Collecting and preparing the documents

Figure 7.4: Pipeline 1 flow chart

Pipeline 1 includes two notebooks:

- `Wikipedia_API.ipynb`, in which we will implement the Wikipedia API to retrieve the URLs of the pages related to the root page of the topic we selected, including the citations for each page. As mentioned, the topic is "marketing" in our case.

- `Knowledge_Graph_Deep_Lake_LlamaIndex_OpenAI_RAG.ipynb`, in which we will implement all three pipelines. In Pipeline 1, it will fetch the URLs provided by the `Wikipedia_API` notebook, clean them, and load and aggregate them for upserting.

We will begin by implementing the Wikipedia API.

Retrieving Wikipedia data and metadata

Let's begin by building a program to interact with the Wikipedia API to retrieve information about a specific topic, tokenize the retrieved text, and manage citations from Wikipedia articles. You may open `Wikipedia_API.ipynb` in the GitHub repository and follow along.

The program begins by installing the `wikipediaapi` library we need:

```
try:
    import wikipediaapi
except:
```

```
!pip install Wikipedia-API==0.6.0
import wikipediaapi
```

> 💡 **Quick tip:** Enhance your coding experience with the **AI Code Explainer** and **Quick Copy** features. Open this book in the next-gen Packt Reader. Click the **Copy** button (**1**) to quickly copy code into your coding environment, or click the **Explain** button (**2**) to get the AI assistant to explain a block of code to you.

```
                                                    Copy        Explain

function calculate(a, b) {                           1            2
  return {sum: a + b};
};
```

🔒 **The next-gen Packt Reader** is included for free with the purchase of this book. Unlock it by scanning the QR code below or visiting `https://www.packtpub.com/unlock/9781836200918`.

The next step is to define the tokenization function that will be called to count the number of tokens of a summary, as shown in the following excerpt:

```
def nb_tokens(text):
    # More sophisticated tokenization which includes punctuation
    tokens = word_tokenize(text)
    return len(tokens)
```

This function takes a string of text as input and returns the number of tokens in the text, using the NLTK library for sophisticated tokenization, including punctuation. Next, to start retrieving data, we need to set up an instance of the Wikipedia API with a specified language and user agent:

```
# Create an instance of the Wikipedia API with a detailed user agent
wiki = wikipediaapi.Wikipedia(
    language='en',
    user_agent='Knowledge/1.0 ([USER AGENT EMAIL)'
)
```

In this case, English was defined with `'en'`, and you must enter the user agent information, such as an email address, for example. We can now define the main topic and filename associated with the Wikipedia page of interest:

```
topic="Marketing"      # topic
filename="Marketing"   # filename for saving the outputs
maxl=100
```

The three parameters defined are:

- `topic`: The topic of the retrieval process
- `filename`: The name of the topic that will customize the files we produce, which can be different from the topic
- `maxl`: The maximum number of URL links of the pages we will retrieve

We now need to retrieve the summary of the specified Wikipedia page, check if the page exists, and print its summary:

```
import textwrap # to wrap the text and display it in paragraphs
page=wiki.page(topic)

if page.exists()==True:
  print("Page - Exists: %s" % page.exists())
  summary=page.summary
  # number of tokens)
  nbt=nb_tokens(summary)
  print("Number of tokens: ",nbt)
  # Use textwrap to wrap the summary text to a specified width, e.g., 70
characters
  wrapped_text = textwrap.fill(summary, width=60)
  # Print the wrapped summary text
  print(wrapped_text)
else:
  print("Page does not exist")
```

The output provides the control information requested:

```
Page - Exists: True
Number of tokens:  229
Marketing is the act of satisfying and retaining customers.
It is one of the primary components of business management
and commerce. Marketing is typically conducted by the seller, typically a
retailer or manufacturer…
```

The information provided shows if we are on the right track or not before running a full search on the main page of the topic:

- Page - Exists: True confirms that the page exists. If not, the print("Page does not exist") message will be displayed.

- Number of tokens: 229 provides us with insights into the size of the content we are retrieving for project management assessments.

- The output of summary=page.summary displays a summary of the page.

In this case, the page exists, fits our topic, and the summary makes sense. Before we continue, we check if we are working on the right page to be sure:

```
print(page.fullurl)
```

The output is correct:

```
https://en.wikipedia.org/wiki/Marketing
```

We are now ready to retrieve the URLs, links, and summaries on the target page:

```
# prompt: read the program up to this cell. Then retrieve all the links for
this page: print the link and a summary of each link.
# Get all the links on the page
links = page.links
# Print the link and a summary of each link
urls = []
counter=0
for link in links:
  try:
    counter+=1
    print(f"Link {counter}: {link}")
    summary = wiki.page(link).summary
    print(f"Link: {link}")
    print(wiki.page(link).fullurl)
    urls.append(wiki.page(link).fullurl)
    print(f"Summary: {summary}")
    if counter>=maxl:
      break
  except page.exists()==False:
    # Ignore pages that don't exist
    pass

print(counter)
print(urls)
```

The function is limited to max1, defined at the beginning of the program. The function will retrieve URL links up to max1 links, or less if the page contains fewer links than the maximum requested. We then check the output before moving on to the next step and generating files:

```
Link 1: 24-hour news cycle
Link: 24-hour news cycle
https://en.wikipedia.org/wiki/24-hour_news_cycle
Summary: The 24-hour news cycle (or 24/7 news cycle) is 24-hour investigation
and reporting of news, concomitant with fast-paced lifestyles…
```

We observe that we have the information we need, and the summaries are acceptable:

*   `Link 1`: The link counter
*   `Link`: The actual link to the page retrieved from the main topic page
*   `Summary`: A summary of the link to the page

The next step is to apply the function we just built to generate the text file containing citations for the links retrieved from a Wikipedia page and their URLs:

```python
from datetime import datetime
# Get all the links on the page
links = page.links


# Prepare a file to store the outputs
fname = filename+"_citations.txt"
with open(fname, "w") as file:
    # Write the citation header
    file.write(f"Citation. In Wikipedia, The Free Encyclopedia. Pages retrieved
from the following Wikipedia contributors on {datetime.now()}\n")
    file.write("Root page: " + page.fullurl + "\n")
    counter = 0
    urls = []…
```

`urls = []` will be appended to have the full list of URLs we need for the final step. The output is a file containing the name of the topic, `datetime`, and the citations beginning with the citation text:

```
Citation. In Wikipedia, The Free Encyclopedia. Pages retrieved from the
following Wikipedia contributors on {datetime.now()}\n")
```

The output, in this case, is a file named `Marketing_citations.txt`. The file was downloaded and uploaded to the `/citations` directory of this chapter's directory in the GitHub repository.

With that, the citations page has been generated, displayed in this notebook, and also saved in the GitHub repository to respect Wikipedia's citation terms. The final step is to generate the file containing the list of URLs we will use to fetch the content of the pages we need. We first display the URLs:

```
urls
```

The output confirms we have the URLs required:

```
['https://en.wikipedia.org/wiki/Marketing',
 'https://en.wikipedia.org/wiki/24-hour_news_cycle',
 'https://en.wikipedia.org/wiki/Account-based_marketing',
 …
```

The URLs are written in a file with the topic as a prefix:

```
# Write URLs to a file
ufname = filename+"_urls.txt"
with open(ufname, 'w') as file:
    for url in urls:
        file.write(url + '\n')
print("URLs have been written to urls.txt")
```

In this case, the output is a file named `Marketing_urls.txt` that contains the URLs of the pages we need to fetch. The file was downloaded and uploaded to the `/citations` directory of the chapter's directory in the GitHub repository.

We are now ready to prepare the data for upsertion.

Preparing the data for upsertion

The URLs provided by the Wikipedia API in the `Wikipedia_API.ipynb` notebook will be processed in the `Knowledge_Graph_ Deep_Lake_LlamaIndex_OpenAI_RAG.ipynb` notebook you can find in the GitHub directory of the chapter. The *Installing the environment* section of this notebook is almost the same section as its equivalent section in *Chapter 2, RAG Embedding Vector Stores with Deep Lake and OpenAI*, and *Chapter 3, Building Index-Based RAG with LlamaIndex, Deep Lake, and OpenAI*. In this chapter, however, the list of URLs was generated by the `Wikipedia_API.ipynb` notebook, and we will retrieve it.

First, go to the *Scenario* section of the notebook to define the strategy of the workflow:

```
#File name for file management
graph_name="Marketing"

# Path for vector store and dataset
db="hub://denis76/marketing01"
vector_store_path = db
dataset_path = db
#if True upserts data; if False, passes upserting and goes to connection
pop_vs=True
# if pop_vs==True, overwrite=True will overwrite dataset, False will append it:
ow=True
```

The parameters will determine the behavior of the three pipelines in the notebook:

- graph_name="Marketing": The prefix (topic) of the files we will read and write.
- db="hub://denis76/marketing01": The name of the Deep Lake vector store. You can choose the name of the dataset you wish.
- vector_store_path = db: The path to the vector store.
- dataset_path = db: The path to the dataset of the vector store.
- pop_vs=True: Activates data insertion if True and deactivates it if False.
- ow=True: Overwrites the existing dataset if True and appends it if False.

Then, we can launch the *Pipeline 1: Collecting and preparing the documents* section of the notebook. The program will download the URL list generated in the previous section of this chapter:

```
# Define your variables
if pop_vs==True:
    directory = "Chapter07/citations"
    file_name = graph_name+"_urls.txt"
    download(directory,file_name)
```

It will then read the file and store the URLs in a list named urls. The rest of the code in the *Pipeline 1: Collecting and preparing the documents* section of this notebook follows the same process as the Deep_Lake_LlamaIndex_OpenAI_RAG.ipynb notebook from *Chapter 3*. In *Chapter 3*, the URLs of the web pages were entered manually in a list.

The code will fetch the content in the list of URLs. The program then cleans and prepares the data to populate the Deep Lake vector store.

Pipeline 2: Creating and populating the Deep Lake vector store

The pipeline in this section of Deep_Lake_LlamaIndex_OpenAI_RAG.ipynb was built with the code of *Pipeline 2* from *Chapter 3*. We can see that by creating pipelines as components, we can rapidly repurpose and adapt them to other applications. Also, Activeloop Deep Lake possesses in-built default chunking, embedding, and upserting functions, making it seamless to integrate various types of unstructured data, as in the case of the Wikipedia documents we are upserting.

The output of the display_record(record_number) function shows how seamless the process is. The output displays the ID and metadata such as the file information, the data collected, the text, and the embedded vector:

```
ID:
['a61734be-fe23-421e-9a8b-db6593c48e08']

Metadata:
file_path: /content/data/24-hour_news_cycle.txt
```

```
file_name: 24-hour_news_cycle.txt
file_type: text/plain
file_size: 2763
creation_date: 2024-07-05
last_modified_date: 2024-07-05

…
Text:
['24hour investigation and reporting of news concomitant with fastpaced
lifestyles This article is about the fastpaced cycle of news media in
technologically advanced societies.

Embedding:
[-0.00040736704249866307, 0.009565318934619427, 0.015906672924757004,
-0.009085721336305141, …]
```

And with that, we have successfully repurposed the *Pipeline 2* component of *Chapter 3* and can now move on and build the graph knowledge index.

Pipeline 3: Knowledge graph index-based RAG

It's time to create a knowledge graph index-based RAG pipeline and interact with it. As illustrated in the following figure, we have a lot of work to do:

Pipeline 3: Knowledge Graph Index-Based RAG
`Knowledge_Graph_Deep_Lake_LlamaIndex_OpenAI_RAG.ipynb`

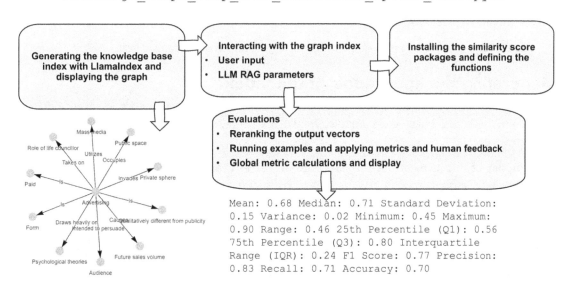

Figure 7.5: Building knowledge graph-index RAG from scratch

In this section, we will:

- Generate the knowledge graph index
- Display the graph
- Define the user prompt
- Define the hyperparameters of LlamaIndex's in-built LLM model
- Install the similarity score packages
- Define the similarity score functions
- Run a sample similarity comparison between the similarity functions
- Re-rank the output vectors of an LLM response
- Run evaluation samples and apply metrics and human feedback scores
- Run metric calculations and display them

Let's go through these steps and begin by generating the knowledge graph index.

Generating the knowledge graph index

We will create a knowledge graph index from a set of documents using the `KnowledgeGraphIndex` class from the `llama_index.core` module. We will also time the index creation process to evaluate performance.

The function begins by recording the start time with `time.time()`. In this case, measuring the time is important because it takes quite some time to create the index:

```
from llama_index.core import KnowledgeGraphIndex
import time
# Start the timer
start_time = time.time()
```

We now create a `KnowledgeGraphIndex` with embeddings using the `from_documents` method. The function uses the following parameters:

- `documents` is the set of documents to index

- `max_triplets_per_chunk` is set to 2, limiting the number of triplets per chunk to optimize memory usage and processing time
- `include_embeddings` is set to `True`, indicating that embeddings should be included

The graph index is thus created in a few lines of code:

```
#graph index with embeddings
graph_index = KnowledgeGraphIndex.from_documents(
    documents,
    max_triplets_per_chunk=2,
    include_embeddings=True,
)
```

The timer is stopped and the creation time is measured:

```
# Stop the timer
end_time = time.time()
# Calculate and print the execution time
elapsed_time = end_time - start_time
print(f"Index creation time: {elapsed_time:.4f} seconds")
print(type(graph_index))
```

The output displays the time:

```
Index creation time: 371.9844 seconds
```

The graph type is displayed:

```
print(type(graph_index))
```

The output confirms the knowledge graph index class:

```
<class 'llama_index.core.indices.knowledge_graph.base.KnowledgeGraphIndex'>
```

We will now set up a query engine for our knowledge graph index and configure it to manage similarity, response temperature, and output length parameters:

```
#similarity_top_k
k=3
#temperature
temp=0.1
#num_output
mt=1024
graph_query_engine = graph_index.as_query_engine(similarity_top_k=k,
temperature=temp, num_output=mt)
```

The parameters will determine the behavior of the query engine:

- k=3 sets the number of top similar results to take into account.
- temp=0.1 sets the temperature parameter, controlling the randomness of the query engine's response generation. The lower it is, the more precise it is; the higher it is, the more creative it is.
- mt=1024 sets the maximum number of tokens for the output, defining the length of the generated responses.

The query engine is then created with the parameters we defined:

```
graph_query_engine = graph_index.as_query_engine(similarity_top_k=k,
temperature=temp, num_output=mt)
```

The graph index and query engine are ready. Let's display the graph.

Displaying the graph

We will create a graph instance, g, with pyvis.network, a Python library used for creating interactive network visualizations. The displayed parameters are similar to the ones we defined in the *Building graphs from trees* section of this chapter:

```
## create graph
from pyvis.network import Network


g = graph_index.get_networkx_graph()
net = Network(notebook=True, cdn_resources="in_line", directed=True)
net.from_nx(g)

# Set node and edge properties: colors and sizes
for node in net.nodes:
    node['color'] = 'lightgray'
    node['size'] = 10


for edge in net.edges:
    edge['color'] = 'black'
    edge['width'] = 1
```

A directed graph has been created, and now we will save it in an HTML file to display it for further use:

```
fgraph="Knowledge_graph_"+ graph_name + ".html"
net.write_html(fgraph)
print(fgraph)
```

The graph_name was defined at the beginning of the notebook, in the *Scenario* section. We will now display the graph in the notebook as an HTML file:

```
from IPython.display import HTML


# Load the HTML content from a file and display it
with open(fgraph, 'r') as file:
    html_content = file.read()
# Display the HTML in the notebook
display(HTML(html_content))
```

You can now download the file to display it in your browser to interact with it. You can also visualize it in the notebook, as shown in the following figure:

Figure 7.6: The knowledge graph

We are all set to interact with the knowledge graph index.

Interacting with the knowledge graph index

Let's now define the functionality we need to execute the query, as we have done in *Chapter 3* in the *Pipeline 3: Index-based RAG* section:

- execute_query is the function we created that will execute the query: response = graph_query_engine.query(user_input). It also measures the time it takes.
- user_query="What is the primary goal of marketing for the consumer market?", which we will use to make the query.

- `response = execute_query(user_query)`, which is encapsulated in the request code and displays the response.

The output provides the best vectors that we created with the Wikipedia data with the time measurement:

```
Query execution time: 2.4789 seconds
The primary goal of marketing for the consumer market is to effectively target
consumers, understand their behavior, preferences, and needs, and ultimately
influence their purchasing decisions.
```

We will now install similarity score packages and define the similarity calculation functions we need.

Installing the similarity score packages and defining the functions

We will first retrieve the Hugging Face token from the **Secrets** tab on Google Colab, where it was stored in the settings of the notebook:

```
from google.colab import userdata
userdata.get('HF_TOKEN')
```

In August 2024, the token is optional for Hugging Face's sentence-transformers. You can ignore the message and comment the code. Next, we install sentence-transformers:

```
!pip install sentence-transformers==3.0.1
```

We then create a cosine similarity function with embeddings:

```
from sklearn.metrics.pairwise import cosine_similarity
from sentence_transformers import SentenceTransformer
model = SentenceTransformer('all-MiniLM-L6-v2')
def calculate_cosine_similarity_with_embeddings(text1, text2):
    embeddings1 = model.encode(text1)
    embeddings2 = model.encode(text2)
    similarity = cosine_similarity([embeddings1], [embeddings2])
    return similarity[0][0]
```

We import the libraries we need:

```
import time
import textwrap
import sys
import io
```

We have a similarity function and can use it for re-ranking.

Re-ranking

In this section, the program re-ranks the response of a query by reordering the top results to select other, possibly better, ones:

- `user_query=" Which experts are often associated with marketing theory?"` represents the query we are making.
- `start_time = time.time()` records the start time for the query execution.
- `response = execute_query(user_query)` executes the query.
- `end_time = time.time()` stops the timer, and the query execution time is displayed.
- `for idx, node_with_score in enumerate(response.source_nodes)` iterates through the response to retrieve all the nodes in the response.
- `similarity_score3=calculate_cosine_similarity_with_embeddings(text1, text2)` calculates the similarity score between the user query and the text in the nodes retrieved from the response. All the comparisons are displayed.
- `best_score=similarity_score3` stores the best similarity score found.
- `print(textwrap.fill(str(best_text), 100))` displays the best re-ranked result.

The initial response for the user_query "Which experts are often associated with marketing theory?" was:

```
Psychologists, cultural anthropologists, and market researchers are often
associated with marketing
theory.
```

The response is acceptable. However, the re-ranked response goes deeper and mentions the names of marketing experts (highlighted in bold font):

```
Best Rank: 2
Best Score: 0.5217772722244263
[…In 1380 the German textile manufacturer Johann Fugger  travelled from
Augsburg to Graben in order to gather information on the international textile
industry… During this period Daniel Defoe  a
London merchant published information on trade and economic resources of
England and Scotland…]
```

The re-ranked response is longer and contains raw document content instead of the summary provided by LlamaIndex's LLM query engine. The original query engine response is better from an LLM perspective. However, it isn't easy to estimate what an end-user will prefer. Some users like short answers, and some like long documents. We can imagine many other ways of re-ranking documents, such as modifying the prompt, adding documents, and deleting documents. We can even decide to fine-tune an LLM, as we will do in *Chapter 9, Empowering AI Models: Fine-Tuning RAG Data and Human Feedback*. We can also introduce human feedback scores as we did in *Chapter 5, Boosting RAG Performance with Expert Human Feedback*, because, in many cases, mathematical metrics will not capture the accuracy of a response (writing fiction, long answers versus short input, and other complex responses). But we need to try anyway!

Let's perform some of the possible metrics for the examples we are going to run.

Example metrics

To evaluate the knowledge graph index's query engine, we will run ten examples and keep track of the scores. rscores keeps track of human feedback scores while scores=[] keeps track of similarity function scores:

```
# create an empty array score human feedback scores:
rscores =[]
# create an empty score for similarity function scores
scores=[]
```

The number of examples can be increased as much as necessary depending on the needs of a project. Each of the ten examples has the same structure:

* user_query, which is the input text for the query engine
* elapsed_time, which is the result of the time measurement of the system's response
* response = execute_query(user_query) executes the query

The user query and output are the same as in the example used for the re-ranking function:

```
Query execution time: 1.9648 seconds
Psychologists, cultural anthropologists, and other experts in behavioral
sciences are often
associated with marketing theory.
```

However, this time, we will run a similarity function and also ask a human for a score:

```
text1=str(response)
text2=user_query
similarity_score3=calculate_cosine_similarity_with_embeddings(text1, text2)
print(f"Cosine Similarity Score with sentence transformer: {similarity_
score3:.3f}")
scores.append(similarity_score3)
human_feedback=0.75
rscores.append(human_feedback)
```

In this function:

* text1 is the query engine's response.
* text2 is the user query.
* similarity_score3 is the cosine similarity score.
* scores.append(similarity_score3) appends the similarity score to scores.

- human_feedback is the human similarity evaluation. We could replace this score with a document as we did in *Chapter 5, Boosting RAG Performance with Expert Human Feedback*, or we could replace the human score with a human text response, which will become the ground truth. In both cases, the similarity score is recalculated with human feedback content.

- rscores.append(human_feedback) appends the human score to rscores.

Let's review a few of the ten examples' outputs and add a comment at the end of each one.

 LLMs are stochastic algorithms. As such, the responses and scores may vary from one run to another.

- Example 1:
  - **User query:** Which experts are often associated with marketing theory?
  - **Response:** Psychologists, cultural anthropologists, and other experts in behavioral sciences are often associated with marketing theory.
  - **Cosine similarity score:** 0.809
  - **Human feedback:** 0.75
  - **Comment:** The response is acceptable, but it could be more specific and mention the names of experts. However, the prompt is ambiguous and only mentions experts in general.

- Example 3:
  - **User query:** What is the difference between B2B and B2C?
  - **Response:** B2B businesses sell products and services to other companies, while B2C businesses sell directly to customers.
  - **Cosine Similarity score:** 0.760
  - **Human feedback:** 0.8
  - **Comment:** The response is precise, but in some cases, users like examples.

- Example 7:
  - **User query:** What commodity programs does the Agricultural Marketing Service (AMS) maintain?
  - **Response:** The **Agricultural Marketing Service (AMS)** maintains programs in five commodity areas: cotton and tobacco, dairy, fruit and vegetable, livestock and seed, and poultry.
  - **Cosine Similarity score:** 0.904
  - **Human feedback:** 0.9

•   **Comment:** This response is accurate and interesting because the information is contained in a page linked to the main page. Thus, this is information from a linked page to the main page. We could ask Wikipedia to search the links of all the linked pages to the main page and go down several levels. However, the main information we are looking for may be diluted in less relevant data. The decision on the scope of the depth of the data depends on the needs of each project.

We will now perform metric calculations on the cosine similarity scores and the human feedback scores.

Metric calculation and display

The cosine similarity scores of the examples are stored in `scores`:

```
print(len(scores), scores)
```

The ten scores are displayed:

```
10 [0.808918, 0.720165, 0.7599532, 0.8513956, 0.5457667, 0.6963912, 0.9036964,
0.44829217, 0.59976315, 0.47448665]
```

We could expand the evaluations to as many other examples, depending on the needs of each project. The human feedback scores for the same examples are stored in `rscores`:

```
print(len(rscores), rscores)
```

The ten human feedback scores are displayed:

```
10 [0.75, 0.5, 0.8, 0.9, 0.65, 0.8, 0.9, 0.2, 0.2, 0.9]
```

We apply metrics to evaluate the responses:

```
mean_score = np.mean(scores)
median_score = np.median(scores)
std_deviation = np.std(scores)
variance = np.var(scores)
min_score = np.min(scores)
max_score = np.max(scores)
range_score = max_score - min_score
percentile_25 = np.percentile(scores, 25)
percentile_75 = np.percentile(scores, 75)
iqr = percentile_75 - percentile_25
```

Each metric can provide several insights. Let's go through each of them and the outputs obtained:

•   **Central tendency (mean, median)** gives us an idea of what a typical score looks like.
•   **Variability (standard deviation, variance, range, IQR)** tells us how spread out the scores are, indicating the consistency or diversity of the data.
•   **Extremes (minimum, maximum)** show the bounds of our dataset.

- **Distribution (percentiles)** provides insights into how scores are distributed across the range of values.

Let's go through these metrics calculated from the cosine similarity scores and the human feedback scores and display their outputs:

1. **Mean (average):**

    - **Definition:** The mean is the sum of all the scores divided by the number of scores.
    - **Purpose:** It gives us the central value of the data, providing an idea of the typical score.
    - **Calculation:**

$$Mean = \frac{\Sigma \, scores}{number \, of \, scores}$$

    - **Output:** `Mean: 0.68`

2. **Median:**

    - **Definition:** The median is the middle value when the scores are ordered from smallest to largest.
    - **Purpose:** It provides the central point of the dataset and is less affected by extreme values (outliers) compared to the mean.
    - **Output:** `Median: 0.71`

3. **Standard deviation:**

    - **Definition:** The standard deviation measures the average amount by which each score differs from the mean.
    - **Purpose:** It gives an idea of how spread out the scores are around the mean. A higher value indicates more variability.
    - **Calculation:**

$$Standard \, Deviation = \sqrt{\frac{\Sigma(score - mean)^2}{number \, of \, scores}}$$

    - **Output:** `Standard Deviation: 0.15`

4. **Variance:**

    - **Definition:** The variance is the square of the standard deviation.
    - **Purpose:** It also measures the spread of the scores, showing how much they vary from the mean.
    - **Output:** `Variance: 0.02`

5. **Minimum:**

   - **Definition:** The minimum is the smallest score in the dataset.
   - **Purpose:** It tells us the lowest value.
   - **Output:** `Minimum: 0.45`

6. **Maximum:**

   - **Definition:** The maximum is the largest score in the dataset.
   - **Purpose:** It tells us the highest value.
   - **Output:** `Maximum: 0.90`

7. **Range:**

   - **Definition:** The range is the difference between the maximum and minimum scores.
   - **Purpose:** It shows the span of the dataset from the lowest to the highest value.
   - **Calculation:**

     Range = Maximum - Minimum

   - **Output:** `Range: 0.46`

8. **25th Percentile (Q1):**

   - **Definition:** The 25th percentile is the value below which 25% of the scores fall.
   - **Purpose:** It provides a point below which a quarter of the data lies.
   - **Output:** `25th Percentile (Q1): 0.56`

9. **75th Percentile (Q3):**

   - **Definition:** The 75th percentile is the value below which 75% of the scores fall.
   - **Purpose:** It gives a point below which three-quarters of the data lies.
   - **Output:** `75th Percentile (Q3): 0.80`

10. **Interquartile Range (IQR):**

    - **Definition:** The IQR is the range between the 25th percentile (Q1) and the 75th percentile (Q3).
    - **Purpose:** It measures the middle 50% of the data, providing a sense of the data's spread without being affected by extreme values.
    - **Calculation:**

      IQR = Q3 – Q1

    - **Output:** `Interquartile Range (IQR): 0.24`

We have built a knowledge-graph-based RAG system, interacted with it, and evaluated it with some examples and metrics. Let's sum up our journey.

Summary

In this chapter, we explored the creation of a scalable knowledge-graph-based RAG system using the Wikipedia API and LlamaIndex. The techniques and tools developed are applicable across various domains, including data management, marketing, and any field requiring organized and accessible data retrieval.

Our journey began with data collection in *Pipeline 1*. This pipeline focused on automating the retrieval of Wikipedia content. Using the Wikipedia API, we built a program to collect metadata and URLs from Wikipedia pages based on a chosen topic, such as marketing. In *Pipeline 2*, we created and populated the Deep Lake vector store. The retrieved data from *Pipeline 1* was embedded and upserted into the Deep Lake vector store. This pipeline highlighted the ease of integrating vast amounts of data into a structured vector store, ready for further processing and querying. Finally, in *Pipeline 3*, we introduced knowledge graph index-based RAG. Using LlamaIndex, we automatically built a knowledge graph index from the embedded data. This index visually mapped out the relationships between different pieces of information, providing a semantic overview of the data. The knowledge graph was then queried using LlamaIndex's built-in language model to generate optimal responses. We also implemented metrics to evaluate the system's performance, ensuring accurate and efficient data retrieval.

By the end of this chapter, we had constructed a comprehensive, automated RAG-driven knowledge graph system capable of collecting, embedding, and querying vast amounts of Wikipedia data with minimal human intervention. This journey showed the power and potential of combining multiple AI tools and models to create an efficient pipeline for data management and retrieval. You are now all set to implement knowledge graph-based RAG systems in real-life projects. In the next chapter, we will learn how to implement dynamic RAG for short-term usage.

Questions

Answer the following questions with yes or no:

1. Does the chapter focus on building a scalable knowledge-graph-based RAG system using the Wikipedia API and LlamaIndex?
2. Is the primary use case discussed in the chapter related to healthcare data management?
3. Does *Pipeline 1* involve collecting and preparing documents from Wikipedia using an API?
4. Is Deep Lake used for creating a relational database in *Pipeline 2*?
5. Does *Pipeline 3* utilize LlamaIndex to build a knowledge graph index?
6. Is the system designed to only handle a single specific topic, such as marketing, without flexibility?
7. Does the chapter describe how to retrieve URLs and metadata from Wikipedia pages?
8. Is a GPU required to run the pipelines described in the chapter?
9. Does the knowledge graph index visually map out relationships between pieces of data?
10. Is human intervention required at every step to query the knowledge graph index?

References

- Wikipedia API GitHub repository: `https://github.com/martin-majlis/Wikipedia-API`
- PyVis Network: *Interactive Network Visualization in Python.*

Further reading

- Hogan, A., Blomqvist, E., Cochez, M., et al. *Knowledge Graphs.* `arXiv:2003.02320`

Learn further in a live workshop with the author

If you are curious how RAG-based systems scale into agentic workflows using context engineering, consider registering for the author's live workshop on 24 January 2026, which extends the ideas discussed here into hands-on practice.

`https://packt.link/gk5Yu`

8

Dynamic RAG with Chroma and Hugging Face Llama

This chapter will take you into the pragmatism of dynamic RAG. In today's rapidly evolving landscape, the ability to make swift, informed decisions is more crucial than ever. Decision-makers across various fields—from healthcare and scientific research to customer service management—increasingly require real-time data that is relevant only within the short period it is needed. A meeting may only require temporary yet highly prepared data. Hence, the concept of data permanence is shifting. Not all information must be stored indefinitely; instead, in many cases, the focus is shifting toward using precise, pertinent data tailored for specific needs at specific times, such as daily briefings or critical meetings.

This chapter introduces an innovative and efficient approach to handling such data through the embedding and creation of temporary Chroma collections. Each morning, a new collection is assembled containing just the necessary data for that day's meetings, effectively avoiding long-term data accumulation and management overhead. This data might include medical reports for a healthcare team discussing patient treatments, customer interactions for service teams strategizing on immediate issues, or the latest scientific research data for researchers making day-to-day experimental decisions. We will then build a Python program to support dynamic and efficient decision-making in daily meetings, applying a methodology using a hard science (any of the natural or physical sciences) dataset for a daily meeting. This approach will highlight the flexibility and efficiency of modern data management. In this case, the team wants to obtain pertinent scientific information without searching the web or interacting with online AI assistants. The constraint is to have a free, open-source assistant that anyone can use, which is why we will use Chroma and Hugging Face resources.

The first step is to create a temporary Chroma collection. We will simulate the processing of a fresh dataset compiled daily, tailored to the specific agenda of upcoming meetings, ensuring relevance and conciseness. In this case, we will download the SciQ dataset from Hugging Face, which contains thousands of crowdsourced science questions, such as those related to physics, chemistry, and biology. Then, the program will embed the relevant data required for the day, guaranteeing that all discussion points are backed by the latest, most relevant data.

A user might choose to run queries before the meetings to confirm their accuracy and alignment with the day's objective. Finally, as meetings progress, any arising questions trigger real-time data retrieval, augmented through **Large Language Model Meta AI (Llama)** technology to generate dynamic flashcards. These flashcards provide quick and precise responses to ensure discussions are both productive and informed. By the end of this chapter, you will have acquired the skills to implement open-source free dynamic RAG in a wide range of domains.

To sum that up, this chapter covers the following topics:

- The architecture of dynamic RAG
- Preparing a dataset for dynamic RAG
- Creating a Chroma collection
- Embedding and upserting data in a Chroma collection
- Batch-querying a collection
- Querying a collection with a user request
- Augmenting the input with the output of a query
- Configuring Hugging Face's framework for Meta Llama
- Generating a response based on the augmented input

Let's begin by going through the architecture of dynamic RAG.

The architecture of dynamic RAG

Imagine you're in a dynamic environment in which information changes daily. Each morning, you gather a fresh batch of 10,000+ questions and validated answers from across the globe. The challenge is to access this information quickly and effectively during meetings without needing long-term storage or complicated infrastructure.

This dynamic RAG method allows us to maintain a lean, responsive system that provides up-to-date information without the burden of ongoing data storage. It's perfect for environments where data relevance is short-lived but critical for decision-making.

We will be applying this to a hard science dataset. However, this dynamic approach isn't limited to our specific example. It has broad applications across various domains, such as:

- **Customer support:** Daily updated FAQs can be accessed in real-time to provide quick responses to customer inquiries.
- **Healthcare:** During meetings, medical teams can use the latest research and patient data to answer complex health-related questions.
- **Finance:** Financial analysts can query the latest market data to make informed decisions on investments and strategies.
- **Education:** Educators can access the latest educational resources and research to answer questions and enhance learning.
- **Tech support:** IT teams can use updated technical documentation to solve issues and guide users effectively.

- **Sales and marketing:** Teams can quickly access the latest product information and market trends to answer client queries and strategize.

This chapter implements one type of a dynamic RAG ecosystem. Your imagination is the limit, so feel free to apply this ecosystem to your own projects in different ways. For now, let's see how the dynamic RAG components fit into the ecosystem we described in *Chapter 1, Why Retrieval Augmented Generation?*, in the *RAG ecosystem* section.

We will streamline the integration and use of dynamic information in real-time decision-making contexts, such as daily meetings, in Python. Here's a breakdown of this innovative strategy for each component and its ecosystem component label:

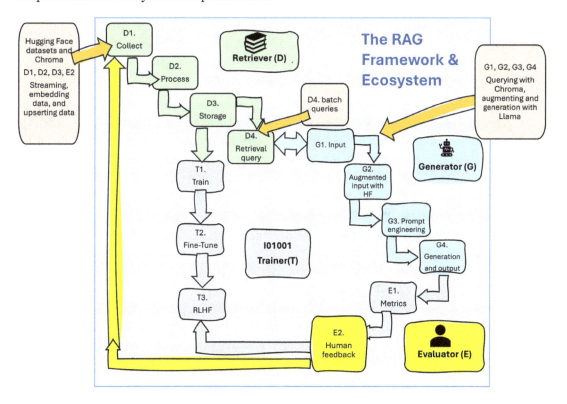

Figure 8.1: The dynamic RAG system

- **Temporary Chroma collection creation (D1, D2, D3, E2):** Every morning, a temporary Chroma collection is set up specifically for that day's meeting. This collection is not meant to be saved post-meeting, serving only the day's immediate needs and ensuring that data does not clutter the system in the long term.

- **Embedding relevant data (D1, D2, D3, E2):** The collection embeds critical data, such as customer support interactions, medical reports, or scientific facts. This embedding process tailors the content specifically to the meeting agenda, ensuring that all pertinent information is at the fingertips of the meeting participants. The data could include human feedback from documents and possibly other generative AI systems.

- **Pre-meeting data validation (D4):** Before the meeting begins, a batch of queries is run against this temporary Chroma collection to ensure that all data is accurate and appropriately aligned with the meeting's objectives, thereby facilitating a smooth and informed discussion.

- **Real-time query handling (G1, G2, G3, G4):** During the meeting, the system is designed to handle spontaneous queries from participants. A single question can trigger the retrieval of specific information, which is then used to augment Llama's input, enabling it to generate flashcards dynamically. These flashcards are utilized to provide concise, accurate responses during the meeting, enhancing the efficiency and productivity of the discussion.

We will be using Chroma, a powerful, open-source, AI-native vector database designed to store, manage, and search embedded vectors in collections. Chroma contains everything we need to start, and we can run it on our machine. It is also very suitable for applications involving LLMs. Chroma collections are thus suitable for a temporary, cost-effective, and real-time RAG system. The dynamic RAG architecture of this chapter implemented with Chroma is innovative and practical. Here are some key points to consider in this fast-moving world:

- **Efficiency and cost-effectiveness:** Using Chroma for temporary storage and Llama for response generation ensures that the system is lightweight and doesn't incur ongoing storage costs. This makes it ideal for environments where data is refreshed frequently and long-term storage isn't necessary. It is very convincing for decision-makers who want lean systems.

- **Flexibility:** The system's ephemeral nature allows for the integration of new data daily, ensuring that the most up-to-date information is always available. This can be particularly valuable in fast-paced environments in which information changes rapidly.

- **Scalability:** The approach is scalable to other similar datasets, provided they can be embedded and queried effectively. This makes it adaptable to various domains beyond the given example. Scaling is not only increasing volumes of data but also the ability to apply a framework to a wide range of domains and situations.

- **User-friendliness:** The system's design is straightforward, making it accessible to users who may not be deeply technical but need reliable answers quickly. This simplicity can enhance user engagement and satisfaction. Making users happy with cost-effective, transparent, and lightweight AI will surely boost their interest in RAG-driven generative AI.

Let's now begin building a dynamic RAG program.

Installing the environment

The environment focuses on open-source and free resources that we can run on our machine or a free Google Colab account. This chapter will run these resources on Google Colab with Hugging Face and Chroma.

We will first install Hugging Face.

Hugging Face

We will implement Hugging Face's open-source resources to download a dataset for the Llama model. Sign up at https://huggingface.co/ to obtain your Hugging Face API token. If you are using Google Colab, you can create a Google Secret in the sidebar and activate it. If so, you can comment the following cell—# Save your Hugging Face token in a secure location:

```
#1.Uncomment the following lines if you want to use Google Drive to retrieve
your token

from google.colab import drive
drive.mount('/content/drive')
f = open("drive/MyDrive/files/hf_token.txt", "r")
access_token=f.readline().strip()
f.close()

#2.Uncomment the following line if you want to enter your HF token manually
#access_token =[YOUR HF_TOKEN]

import os
os.environ['HF_TOKEN'] = access_token
```

The program first retrieves the Hugging Face API token. Make sure to store it in a safe place. You can choose to use Google Drive or enter it manually. Up to now, the installation seems to have run smoothly. We now install datasets:

```
!pip install datasets==2.20.0
```

However, there are conflicts, such as pyarrow, with Google Colab's pre-installed version, which is more recent. These conflicts between fast-moving packages are frequent. When Hugging Face updates its packages, this conflict will not appear anymore. But other conflicts may appear. This conflict will not stop us from downloading datasets. If it did, we would have to uninstall Google Colab packages and reinstall pyarrow, but other dependencies may possibly create issues. We must accept these challenges, as explained in the *Setting up the environment* section in *Chapter 2, RAG Embedding Vector Stores with Deep Lake and OpenAI.*

We will now install Hugging Face's transformers package:

```
!pip install transformers==4.41.2
```

We also install accelerate to run PyTorch packages on GPUs, which is highly recommended for this notebook, among other features, such as mixed precision and accelerated processing times:

```
!pip install accelerate==0.31.0
```

Finally, we will initialize `meta-llama/Llama-2-7b-chat-hf` as the tokenizer and chat model interactions. Llama is a series of transformer-based language models developed by Meta AI (formerly Facebook AI) that we can access through Hugging Face:

```
from transformers import AutoTokenizer
import tranformers
import torch
model = "meta-llama/Llama-2-7b-chat-hf"
tokenizer = AutoTokenizer.from_pretrained(model)
```

We access the model through Hugging Face's pipeline:

```
pipeline = transformers.pipeline(
    "text-generation",
    model=model,
    torch_dtype=torch.float16,
    device_map="auto",
)
```

Let's go through the pipeline:

- `transformers.pipeline` is the function used to create a pipeline for text generation. This pipeline abstracts away much of the complexity we must avoid in this dynamic RAG ecosystem.
- `text-generation` specifies the type of task the pipeline is set up for. In this case, we want text generation.
- `model` specifies the model we selected.
- `torch_dtype=torch.float16` sets the data type for PyTorch tensors to `float16`. This is a key factor for dynamic RAG, which reduces memory consumption and can speed up computation, particularly on GPUs that support half-precision computations. Half-precision computations use 16 bits: half of the standard 32-bit precision, for faster, lighter processing. This is exactly what we need.
- `device_map="auto"` instructs the pipeline to automatically determine the best device to run the model on (CPU, GPU, multi-GPU, etc.). This parameter is particularly important for optimizing performance and automatically distributing the model's layers across available devices (like GPUs) in the most efficient manner possible. If multiple GPUs are available, it will distribute the load across them to maximize parallel processing. If you have access to a GPU, activate it to speed up the configuration of this pipeline.

Hugging Face is ready; Chroma is required next.

Chroma

The following line installs Chroma, our open-source vector database:

```
!pip install chromadb==0.5.3
```

Take a close look at the following excerpt output, which displays the packages installed and, in particular, **Open Neural Network Exchange (ONNX)**:

```
Successfully installed asgiref-3…onnxruntime-1.18.0…
```

ONNX (https://onnxruntime.ai/) is a key component in this chapter's dynamic RAG scenario because it is fully integrated with Chroma. ONNX is a standard format for representing **machine learning (ML)** models designed to enable models to be used across different frameworks and hardware without being locked into one ecosystem.

We will be using ONNX Runtime, which is a performance-focused engine for running ONNX models. It acts as a cross-platform accelerator for ML models, providing a flexible interface that allows integration with hardware-specific libraries. This makes it possible to optimize the models for various hardware configurations (CPUs, GPUs, and other accelerators). As for Hugging Face, it is recommended to activate a GPU if you have access to one for the program in this chapter. Also, we will select a model included within ONNX Runtime installation packages.

We have now installed the Hugging Face and Chroma resources we need, including ONNX Runtime. Hugging Face's framework is used throughout the model life cycle, from accessing and deploying pre-trained models to training and fine-tuning them within its ecosystem. ONNX, among its many features, can intervene in the post-training phase to ensure a model's compatibility and efficient execution across different hardware and software setups. Models might be developed and fine-tuned using Hugging Face's tools and then converted to the ONNX format for broad, optimized deployment using ONNX Runtime.

We will now use spaCy to compute the accuracy between the response we obtain when querying our vector store and the original completion text. The following command installs a medium-sized English language model from spaCy, tailored for general NLP tasks:

```
!python -m spacy download en_core_web_md
```

This model, labeled en_core_web_md, originates from web text in English and is balanced for speed and accuracy, which we need for dynamic RAG. It is efficient for computing text similarity. You may need to restart the session once the package is installed.

We have now successfully installed the open-source, optimized, cost-effective resources we need for dynamic RAG and are ready to start running the program's core.

Activating session time

When working in real-life dynamic RAG projects, such as in this scenario, time is essential! For example, if the daily decision-making meeting is at 10 a.m., the RAG preparation team might have to start preparing for this meeting at 8 a.m. to gather the data online, in processed company data batches, or in any other way necessary for the meeting's goal.

First, activate a GPU if one is available. On Google Colab, for example, go to **Runtime** | **Change runtime type** and select a GPU if possible and available. If not, the notebook will take a bit longer but will run on a CPU. Then, go through each section in this chapter, running the notebook cell by cell to understand the process in depth.

The following code activates a measure of the session time once the environment is installed all the way to the end of the notebook:

```
# Start timing before the request
session_start_time = time.time()
```

Finally, restart the session, go to **Runtime** again, and click on **Run all**. Once the program is finished, go to **Total session time**, the last section of the notebook. You will have an estimate of how long it takes for a preparation run. With the time left before a daily meeting, you can tweak the data, queries, and model parameters for your needs a few times.

This on-the-fly dynamic RAG approach will make any team that has these skills a precious asset in this fast-moving world. We will start the core of the program by downloading and preparing the dataset.

Downloading and preparing the dataset

We will use the SciQ dataset created by Welbl, Liu, and Gardner (2017) with a method for generating high-quality, domain-specific multiple-choice science questions via *crowdsourcing*. The SciQ dataset consists of 13,679 multiple-choice questions crafted to aid the training of NLP models for science exams. The creation process involves two main steps: selecting relevant passages and generating questions with plausible distractors.

In the context of using this dataset for an augmented generation of questions through a Chroma collection, we will implement the question, correct_answer, and support columns. The dataset also contains distractor columns with wrong answers, which we will drop.

We will integrate the prepared dataset into a retrieval system that utilizes query augmentation techniques to enhance the retrieval of relevant questions based on specific scientific topics or question formats for Hugging Face's Llama model. This will allow for the dynamic generation of augmented, real-time completions for Llama, as implemented in the chapter's program. The program loads the training data from the sciq dataset:

```
# Import required libraries
from datasets import load_dataset
import pandas as pd


# Load the SciQ dataset from HuggingFace
dataset = load_dataset("sciq", split="train")
```

The dataset is filtered to detect the non-empty support and correct_answer columns:

```
# Filter the dataset to include only questions with support and correct answer
filtered_dataset = dataset.filter(lambda x: x["support"] != "" and x["correct_
answer"] != "")
```

We will now display the number of rows filtered:

```
# Print the number of questions with support
print("Number of questions with support: ", len(filtered_dataset))
```

The output shows that we have 10,481 documents:

```
Number of questions with support:  10481
```

We need to clean the DataFrame to focus on the columns we need. Let's drop the distractors (wrong answers to the questions):

```
# Convert the filtered dataset to a pandas DataFrame
df = pd.DataFrame(filtered_dataset)

# Columns to drop
columns_to_drop = ['distractor3', 'distractor1', 'distractor2']

# Dropping the columns from the DataFrame
df.drop(columns=columns_to_drop, inplace=True)
```

We have the correct answer and the support content that we will now merge:

```
# Create a new column 'completion' by merging 'correct_answer' and 'support'
df['completion'] = df['correct_answer'] + " because " + df['support']

# Ensure no NaN values are in the 'completion' column
df.dropna(subset=['completion'], inplace=True)
df
```

The output shows the columns we need to prepare the data for retrieval in the completion columns, as shown in the excerpt of the DataFrame for a completion field in which aerobic is the correct answer because it is the connector and the rest of the text is the support content for the correct answer:

```
aerobic because "Cardio" has become slang for aerobic exercise that raises your
heart rate for an extended amount of time. Cardio can include biking, running,
or swimming. Can you guess one of the main organs of the cardiovascular system?
Yes, your heart.
```

The program now displays the shape of the DataFrame:

```
df.shape
```

The output shows we still have all the initial lines and four columns:

```
(10481, 4)
```

The following code will display the names of the columns:

```
# Assuming 'df' is your DataFrame
print(df.columns)
```

As a result, the output displays the four columns we need:

```
Index(['question', 'correct_answer', 'support', 'completion'], dtype='object')
```

The data is now ready to be embedded and upserted.

Embedding and upserting the data in a Chroma collection

We will begin by creating the Chroma client and defining a collection name:

```
# Import Chroma and instantiate a client. The default Chroma client is
ephemeral, meaning it will not save to disk.
import chromadb
client = chromadb.Client()
collection_name="sciq_supports6"
```

Before creating the collection and upserting the data to the collection, we need to verify whether the collection already exists or not:

```
# List all collections
collections = client.list_collections()


# Check if the specific collection exists
collection_exists = any(collection.name == collection_name for collection in
collections)
print("Collection exists:", collection_exists)
```

The output will return True if the collection exists and False if it doesn't:

```
Collection exists: False
```

If the collection doesn't exist, we will create a collection with collection_name defined earlier:

```
# Create a new Chroma collection to store the supporting evidence. We don't
need to specify an embedding function, and the default will be used.
if collection_exists!=True:
  collection = client.create_collection(collection_name)
else:
  print("Collection ", collection_name," exists:", collection_exists)
```

Let's peek into the structure of the dictionary of the collection we created:

```
#Printing the dictionary
results = collection.get()
for result in results:
    print(result)  # This will print the dictionary for each item
```

The output displays the dictionary of each item of the collection:

```
ids
embeddings
metadatas
documents
uris
data
included
```

Let's briefly go through the three key fields for our scenario:

- `ids`: This field represents the unique identifiers for each item in the collection.
- `embeddings`: Embeddings are the embedded vectors of the documents.
- `documents`: This refers to the `completion` column in which we merged the correct answer and the support content.

We now need a lightweight rapid LLM model for our dynamic RAG environment.

Selecting a model

Chroma will initialize a default model, which can be `all-MiniLM-L6-v2`. However, let's make sure we are using this model and initialize it:

```
model_name = "all-MiniLM-L6-v2"  # The name of the model to use for embedding
and querying
```

The `all-MiniLM-L6-v2` model was designed with an optimal, enhanced method by Wang et al. (2021) for model compression, focusing on distilling self-attention relationships between components of transformer models. This approach is flexible in the number of attention heads between teacher and student models, improving compression efficiency. The model is fully integrated into Chroma with ONNX, as explained in the *Installing the environment* section of this chapter.

The magic of this `MiniLM` model is based on compression and knowledge distillation through a teacher model and the student model:

- **Teacher model:** This is the original, typically larger and more complex model such as BERT, RoBERTa, and XLM-R, in our case, that has been pre-trained on a comprehensive dataset. The teacher model possesses high accuracy and a deep understanding of the tasks it has been trained on. It serves as the source of knowledge that we aim to transfer.
- **Student model:** This is our smaller, less complex model, `all-MiniLM-L6-v2`, which is trained to mimic the teacher model's behavior, which will prove very effective for our dynamic RAG architecture. The goal is to have the student model replicate the performance of the teacher model as closely as possible but with significantly fewer parameters or computational expense.

In our case, `all-MiniLM-L6-v2` will accelerate the embedding and querying process. We can see that in the age of superhuman LLM models, such as GPT-4o, we can perform daily tasks with smaller compressed and distilled models. Let's embed the data next.

Embedding and storing the completions

Embedding and upserting data in a Chroma collection is seamless and concise. In this scenario, we'll embed and upsert the whole df completions in a completion_list extracted from our df dataset:

```
ldf=len(df)
nb=ldf  # number of questions to embed and store

import time
start_time = time.time()  # Start timing before the request

# Convert Series to list of strings
completion_list = df["completion"][:nb].astype(str).tolist()
```

We use the collection_exists status we defined when creating the collection to avoid loading the data twice. In this scenario, the collection is temporary; we just want to load it once and use it once. If you try to load the data in this temporary scenario a second time, you will get warnings. However, you can modify the code if you wish to try different datasets and methods, such as preparing a prototype at full speed for another project.

In any case, in this scenario, we first check if the collection exists and then upsert the ids and documents in the complete_list and store the type of data, which is completion, in the metadatas field:

```
# Avoiding trying to load data twice in this one run dynamic RAG notebook
if collection_exists!=True:
  # Embed and store the first nb supports for this demo
  collection.add(
      ids=[str(i) for i in range(0, nb)],  # IDs are just strings
      documents=completion_list,
      metadatas=[{"type": "completion"} for _ in range(0, nb)],
  )
```

Finally, we measure the response time:

```
response_time = time.time() - start_time  # Measure response time
print(f"Response Time: {response_time:.2f} seconds")  # Print response time
```

The output shows that, in this case, Chroma activated the default model through onnx, as explained in the introduction of this section and also in the *Installing the environment* section of this chapter:

```
/root/.cache/chroma/onnx_models/all-MiniLM-L6-v2/onnx.tar.gz: 100%|        |
79.3M/79.3M [00:02<00:00, 31.7MiB/s]
```

The output also shows that the processing time for 10,000+ documents is satisfactory:

```
Response Time: 234.25 seconds
```

The response time might vary and depends on whether you are using a GPU. When using an accessible GPU, the time fits the needs required for dynamic RAG scenarios.

With that, the Chroma vector store is now populated. Let's take a peek at the embeddings.

Displaying the embeddings

The program now fetches the embeddings and displays the first one:

```
# Fetch the collection with embeddings included
result = collection.get(include=['embeddings'])

# Extract the first embedding from the result
first_embedding = result['embeddings'][0]

# If you need to work with the length or manipulate the first embedding:
embedding_length = len(first_embedding)

print("First embedding:", first_embedding)
print("Embedding length:", embedding_length)
```

The output shows that our completions have been vectorized, as we can see in the first embedding:

```
First embedding: [0.03689068928360939, -0.05881563201546669,
-0.04818134009838104,…
```

The output also displays the embedding length, which is interesting:

```
Embedding length: 384
```

The all-MiniLM-L6-v2 model reduces the complexity of text data by mapping sentences and paragraphs into a 384-dimensional space. This is significantly lower than the typical dimensionality of one-hot encoded vectors, such as the 1,526 dimensions of the OpenAI text-embedding-ada-002. This shows that all-MiniLM-L6-v2 uses dense vectors, which use all dimensions of the vector space to encode information to produce nuanced semantic relationships between different documents as opposed to sparse vectors.

Sparse vector models, such as the **bag-of-words** (BoW) model, can be effective in some cases. However, their main limitation is that they don't capture the order of words or the context around them, which can be crucial for understanding the meaning of text when training LLMs.

We have now embedded the documents into dense vectors in a smaller dimensional space than full-blown LLMs and will produce satisfactory results.

Querying the collection

The code in this section executes a query against the Chroma vector store using its integrated semantic search functionality. It queries the vector representations of all the vectors in the Chroma collection questions in the initial dataset:

```
dataset["question"][:nbq].
```

The query requests one most relevant or similar document for each question with n_results=1, which you can modify if you wish.

Each question text is converted into a vector. Then, Chroma runs a vector similarity search by comparing the embedded vectors against our database of document vectors to find the closest match based on vector similarity:

```
import time
start_time = time.time()  # Start timing before the request


# number of retrievals to write
results = collection.query(
    query_texts=df["question"][:nb],
    n_results=1)


response_time = time.time() - start_time  # Measure response time
print(f"Response Time: {response_time:.2f} seconds")  # Print response time
```

The output displays a satisfactory response time for the 10,000+ queries:

```
Response Time: 199.34 seconds
```

We will now analyze the 10,000+ queries. We will use spaCy to evaluate a query's result and compare it with the original completion. We first load the spaCy model we installed in the *Installing the environment* section of this chapter:

```
import spacy
import numpy as np


# Load the pre-trained spaCy language model
nlp = spacy.load('en_core_web_md')  # Ensure that you've installed this model
with 'python -m spacy download en_core_web_md'
```

The program then creates a similarity function that takes two arguments (the original completion, text1, and the retrieved text, text2) and returns the similarity value:

```
def simple_text_similarity(text1, text2):
    # Convert the texts into spaCy document objects
    doc1 = nlp(text1)
```

```
    doc2 = nlp(text2)

    # Get the vectors for each document
    vector1 = doc1.vector
    vector2 = doc2.vector

    # Compute the cosine similarity between the two vectors
    # Check for zero vectors to avoid division by zero
    if np.linalg.norm(vector1) == 0 or np.linalg.norm(vector2) == 0:
        return 0.0  # Return zero if one of the texts does not have a vector
representation
    else:
        similarity = np.dot(vector1, vector2) / (np.linalg.norm(vector1) *
np.linalg.norm(vector2))
        return similarity
```

We will now perform a full validation run on the 10,000 queries. As can be seen in the following code block, the validation begins by defining the variables we will need:

- nbqd to only display the first 100 and last 100 results.
- acc_counter measures the results with a similarity score superior to 0.5, which you can modify to fit your needs.
- display_counter to count the number of results we have displayed:

```
nbqd = 100  # the number of responses to display, supposing there are more than
100 records

# Print the question, the original completion, the retrieved document, and
compare them
acc_counter=0
display_counter=0
```

The program goes through nb results, which, in our case, is the total length of our dataset:

```
for i, q in enumerate(df['question'][:nb]):
    original_completion = df['completion'][i]  # Access the original completion
for the question
    retrieved_document = results['documents'][i][0]  # Retrieve the
corresponding document
    similarity_score = simple_text_similarity(original_completion, retrieved_
document)
```

The code accesses the original completion and stores it in original_completion. Then, it retrieves the result and stores it in retrieved_document. Finally, it calls the similarity function we defined, simple_text_similarity. The original completion and the retrieved document store the similarity score in similarity_score.

Now, we introduce an accuracy metric. In this scenario, the threshold of the similarity score is set to `0.7`, which is reasonable:

```
if similarity_score > 0.7:
    acc_counter+=1
```

If `similarity_score > 0.7`, then the accuracy counter, `acc_counter`, is incremented. The display counter, `display_counter`, is also incremented to only the first and last `nbqd` (maximum results to display) defined at the beginning of this function:

```
display_counter+=1
if display_counter<=nbqd or display_counter>nb-nbqd:
```

The information displayed provides insights into the performance of the system:

```
print(i," ", f"Question: {q}")
print(f"Retrieved document: {retrieved_document}")
print(f"Original completion: {original_completion}")
print(f"Similarity Score: {similarity_score:.2f}")
print()  # Blank line for better readability between entries
```

The output displays four key variables:

* `{q}` is the question asked, the query.
* `{retrieved_document}` is the document retrieved.
* `{original_completion}` is the original document in the dataset.
* `{similarity_score:.2f}` is the similarity score between the original document and the document retrieved to measure the performance of each response.

The first output provides the information required for a human observer to control the result of the query and trace it back to the source.

The first part of the output is the question, the query:

```
Question: What type of organism is commonly used in preparation of foods such
as cheese and yogurt?
```

The second part of the output is the retrieved document:

```
Retrieved document: lactic acid because Bacteria can be used to make cheese
from milk. The bacteria turn the milk sugars into lactic acid. The acid is
what causes the milk to curdle to form cheese. Bacteria are also involved in
producing other foods. Yogurt is made by using bacteria to ferment milk (
Figure below ). Fermenting cabbage with bacteria produces sauerkraut.
```

The third part of the output is the original completion. In this case, we can see that the retrieved document provides relevant information but not the exact original completion:

```
Original completion: mesophilic organisms because Mesophiles grow best in
moderate temperature, typically between 25°C and 40°C (77°F and 104°F).
Mesophiles are often found living in or on the bodies of humans or other
animals. The optimal growth temperature of many pathogenic mesophiles is 37°C
(98°F), the normal human body temperature. Mesophilic organisms have important
uses in food preparation, including cheese, yogurt, beer and wine.
```

Finally, the output displays the similarity score calculated by spaCy:

```
Similarity Score: 0.73
```

The score shows that although the original completion was not selected, the completion selected is relevant.

When all the results have been analyzed, the program calculates the accuracy obtained for the 10,000+ queries:

```
if nb>0:
  acc=acc_counter/nb
```

The calculation is based on the following:

- Acc is the overall accuracy obtained
- acc_counter is the total of Similarity scores > 0.7
- nb is the number of queries. In this case, nb=len(df)
- acc=acc_counter/nb calculates the overall accuracy of all the results

The code then displays the number of documents measured and the overall similarity score:

```
    print(f"Number of documents: {nb:.2f}")
    print(f"Overall similarity score: {acc:.2f}")
```

The output shows that all the questions returned relevant results:

```
Number of documents: 10481.00
Overall similarity score: 1.00
```

This satisfactory overall similarity score shows that the system works in a closed environment. But we need to go further and see what happens in the open environment of heated discussions in a meeting!

Prompt and retrieval

This section is the one to use during real-time querying meetings. You can adapt the interface to your needs. We'll focus on functionality.

Let's look at the first prompt:

```
# initial question
prompt = "Millions of years ago, plants used energy from the sun to form what?"
# variant 1 similar
```

```
#prompt = "Eons ago, plants used energy from the sun to form what?"
# variant 2 divergent
#prompt = "Eons ago, plants used sun energy to form what?"
```

You will notice that there are two commented variants under the first prompt. Let's clarify this:

- `initial question` is the exact text that comes from the initial dataset. It isn't likely that an attendee in the meeting or a user will ask the question that way. But we can use it to verify if the system is working.
- `variant 1` is similar to the initial question and could be asked.
- `variant 2` diverges and may prove challenging.

We will select `variant 1` for this section and we should obtain a satisfactory result.

We can see that, as for all AI programs, human control is mandatory! The more `variant 2` diverges with spontaneous questions, the more challenging it becomes for the system to remain stable and respond as we expect. This limit explains why, even if a dynamic RAG system can adapt rapidly, designing a solid system will require careful and continual improvements.

If we query the collection as we did in the previous section with one prompt only this time, we will obtain a response rapidly:

```python
import time
import textwrap


# Start timing before the request
start_time = time.time()


# Query the collection using the prompt
results = collection.query(
    query_texts=[prompt],  # Use the prompt in a list as expected by the query
method
    n_results=1  # Number of results to retrieve
)


# Measure response time
response_time = time.time() - start_time


# Print response time
print(f"Response Time: {response_time:.2f} seconds\n")


# Check if documents are retrieved
if results['documents'] and len(results['documents'][0]) > 0:
    # Use textwrap to format the output for better readability
```

```
    wrapped_question = textwrap.fill(prompt, width=70)  # Wrap text at 70
characters
    wrapped_document = textwrap.fill(results['documents'][0][0], width=70)

    # Print formatted results
    print(f"Question: {wrapped_question}")
    print("\n")
    print(f"Retrieved document: {wrapped_document}")
    print()
else:
    print("No documents retrieved."
```

The response time is rapid:

```
Response Time: 0.03 seconds
```

The output shows that the retrieved document is relevant:

```
Response Time: 0.03 seconds

Question: Millions of years ago, plants used energy from the sun to form what?

Retrieved document: chloroplasts because When ancient plants underwent
photosynthesis,
they changed energy in sunlight to stored chemical energy in food. The
plants used the food and so did the organisms that ate the plants.
After the plants and other organisms died, their remains gradually
changed to fossil fuels as they were covered and compressed by layers
of sediments. Petroleum and natural gas formed from ocean organisms
and are found together. Coal formed from giant tree ferns and other
swamp plants.
```

We have successfully retrieved the result of our query. This semantic vector search might even be enough if the attendees of the meeting are satisfied with it. You will always have time to improve the configuration of RAG with Llama.

Hugging Face Llama will now take this response and write a brief NLP summary.

RAG with Llama

We initialized `meta-llama/Llama-2-7b-chat-hf` in the *Installing the environment* section. We must now create a function to configure Llama 2's behavior:

```
def LLaMA2(prompt):
    sequences = pipeline(
        prompt,
```

```
        do_sample=True,
        top_k=10,
        num_return_sequences=1,
        eos_token_id=tokenizer.eos_token_id,
        max_new_tokens=100, # Control the output length more granularly
        temperature=0.5,  # Slightly higher for more diversity
        repetition_penalty=2.0,  # Adjust based on experimentation
        truncation=True
    )
    return sequences
```

You can tweak each parameter to your expectations:

- prompt: The input text that the model uses to generate the output. It's the starting point for the model's response.

- do_sample: A Boolean value (True or False). When set to True, it enables stochastic sampling, meaning the model will pick tokens randomly based on their probability distribution, allowing for more varied outputs.

- top_k: This parameter limits the number of highest-probability vocabulary tokens to consider when selecting tokens in the sampling process. Setting it to 10 means the model will choose from the top 10 most likely next tokens.

- num_return_sequences: Specifies the number of independently generated responses to return. Here, it is set to 1, meaning the function will return one sequence for each prompt.

- eos_token_id: This token marks the end of a sequence in tokenized form. Once it is generated, the model stops generating further tokens. The end-of-sequence token is an id that points to Llama's eos_token.

- max_new_tokens: Limits the number of new tokens the model can generate. Set to 100 here, it constrains the output to a maximum length of 100 tokens beyond the input prompt length.

- temperature: This controls randomness in the sampling process. A temperature of 0.5 makes the model's responses less random and more focused than a higher temperature but still allows for some diversity.

- repetition_penalty: A modifier that discourages the model from repeating the same token. A penalty of 2.0 means any token already used is less likely to be chosen again, promoting more diverse and less repetitive text.

- truncation: When enabled, it ensures the output does not exceed the maximum length specified by max_new_tokens by cutting off excess tokens.

The prompt will contain the instruction for Llama in iprompt and the result obtained in the *Prompt and retrieval* section of the notebook. The result is appended to iprompt:

```
iprompt='Read the following input and write a summary for beginners.'
lprompt=iprompt + " " + results['documents'][0][0]
```

The augmented input for the Llama call is lprompt. The code will measure the time it takes and make the completion request:

```
import time
start_time = time.time()  # Start timing before the request


response=LLaMA2(lprompt)
```

We now retrieve the generated text from the response and display the time it took for Llama to respond:

```
for seq in response:
    generated_part = seq['generated_text'].replace(iprompt, '')  # Remove the
input part from the output


response_time = time.time() - start_time  # Measure response time
print(f"Response Time: {response_time:.2f} seconds")  # Print response timeLe
```

The output shows that Llama returned the completion in a reasonable time:

```
Response Time: 5.91 seconds
```

Let's wrap the response in a nice format to display it:

```
wrapped_response = textwrap.fill(response[0]['generated_text'], width=70)
print(wrapped_response)
```

The output displays a technically reasonable completion:

```
chloroplasts because When ancient plants underwent photosynthesis,
they changed energy in sunlight to stored chemical energy in food. The
plants used the food and so did the organisms that ate the plants.
After the plants and other organisms died, their remains gradually
changed to fossil fuels as they were covered and compressed by layers
of sediments. Petroleum and natural gas formed from ocean organisms
and are found together. Coal formed from giant tree ferns and other
swamp plants. Natural Gas: 10% methane (CH4) - mostly derived from
anaerobic decomposition or fermentation processes involving
microorganism such As those present In wetlands; also contains smaller
amounts Of ethene(C2H6), propiene/propadiene/( C3 H5-7). This is where
most petrol comes frm! But there're more complex hydrocarbons like
pentanes & hexans too which can come
```

The summary produced by Llama is technically acceptable. To obtain another, possibly better result, as long as the session is not closed, the user can run a query and an augmented generation several times with different Llama parameters.

You can even try another LLM. Dynamic RAG doesn't necessarily have to be 100% open-source. If necessary, we must be pragmatic and introduce whatever it takes. For example, the following prompt was submitted to ChatGPT with GPT-4o, which is the result of the query we used for Llama:

```
Write a nice summary with this text: Question: Millions of years ago, plants
used energy from the sun to form what?


Retrieved document: chloroplasts because When ancient plants underwent
photosynthesis,
they changed energy in sunlight to stored chemical energy in food. The plants
used the food and so did the organisms that ate the plants. After the plants
and other organisms died, their remains gradually
changed to fossil fuels as they were covered and compressed by layers of
sediments. Petroleum and natural gas formed from ocean organisms and are found
together. Coal formed from giant tree ferns and other swamp plants.
```

The output of OpenAI GPT-4o surpasses Llama 2 in this case and produces a satisfactory output:

```
Millions of years ago, plants harnessed energy from the sun through
photosynthesis to produce food, storing chemical energy. This energy was vital
for the plants themselves and for the organisms that consumed them. Over time,
the remains of these plants and animals, buried under sediment, transformed
into fossil fuels. Ocean organisms' remains contributed to the formation of
petroleum and natural gas, often found together, while the remains of giant
tree ferns and swamp plants formed coal.
```

If necessary, you can replace `meta-llama/Llama-2-7b-chat-hf` with GPT-4o, as implemented in *Chapter 4, Multimodal Modular RAG for Drone Technology*, and configure it to obtain this level of output. The only rule in dynamic RAG is performance. With that, we've seen that there are many ways to implement dynamic RAG.

Once the session is over, we can delete it.

Deleting the collection

You can manually delete the collection with the following code:

```
#client.delete_collection(collection_name)
```

You can also close the session to delete the temporary dynamic RAG collection created. We can check and see whether the collection we created, `collection_name`, still exists or not:

```
# List all collections
collections = client.list_collections()

# Check if the specific collection exists
```

```
collection_exists = any(collection.name == collection_name for collection in
collections)
print("Collection exists:", collection_exists)
```

If we are still working on a collection in a session, the response will be True:

```
Collection exists: True
```

If we delete the collection with code or by closing the session, the response will be False. Let's take a look at the total session time.

Total session time

The following code measures the time between the beginning of the session and immediately after the *Installing the environment* section:

```
end_time = time.time() - session_start_time  # Measure response time
print(f"Session preparation time: {response_time:.2f} seconds")  # Print
response time
```

The output can have two meanings:

- It can measure the time we worked on the preparation of the dynamic RAG scenario with the daily dataset for the Chroma collection, querying, and summarizing by Llama.
- It can measure the time it took to run the whole notebook without intervening at all.

In this case, the session time is the result of a full run with no human intervention:

```
Session preparation time: 780.35 seconds
```

The whole process takes less than 15 minutes, which fits the constraints of the preparation time in a dynamic RAG scenario. It leaves room for a few runs to tweak the system before the meeting. With that, we have successfully walked through a dynamic RAG process and will now summarize our journey.

Summary

In a fast-evolving world, gathering information rapidly for decision-making provides a competitive advantage. Dynamic RAG is one way to bring AI into meeting rooms with rapid and cost-effective AI. We built a system that simulated the need to obtain answers to hard science questions in a daily meeting. After installing and analyzing the environment, we downloaded and prepared the SciQ dataset, a science question-and-answer dataset, to simulate a daily meeting during which hard science questions would be asked. The attendees don't want to spend their time searching the web and wasting their time when decisions must be made. This could be for a marketing campaign, fact-checking an article, or any other situation in which hard science knowledge is required.

We created a Chroma collection vector store. We then embedded 10,000+ documents and inserted data and vectors into the Chroma vector store on our machine with `all-MiniLM-L6-v2`. The process proved cost-effective and sufficiently rapid. The collection was created locally, so there is no storage cost. The collection is temporary, so there is no useless space usage or cluttering. We then queried the collection to measure the accuracy of the system we set up. The results were satisfactory, so we processed the full dataset to confirm. Finally, we created the functionality for a user prompt and query function to use in real time during a meeting. The result of the query augmented the user's input for `meta-llama/Llama-2-7b-chat-hf`, which transformed the query into a short summary.

The dynamic RAG example we implemented would require more work before being released into production. However, it provides a path to open-source, lightweight, RAG-driven generative AI for rapid data collection, embedding, and querying. If we need to store the retrieval data and don't want to create large vector stores, we can integrate our datasets in an OpenAI GPT-4o-mini model, for example, through fine-tuning, as we will see in the next chapter.

Questions

Answer the following questions with *Yes* or *No*:

1. Does the script ensure that the Hugging Face API token is never hardcoded directly into the notebook for security reasons?
2. In the chapter's program, is the `accelerate` library used here to facilitate the deployment of ML models on cloud-based platforms?
3. Is user authentication separate from the API token required to access the Chroma database in this script?
4. Does the notebook use Chroma for temporary storage of vectors during the dynamic retrieval process?
5. Is the notebook configured to use real-time acceleration of queries through GPU optimization?
6. Can this notebook's session time measurements help in optimizing the dynamic RAG process?
7. Does the script demonstrate Chroma's capability to integrate with ML models for enhanced retrieval performance?
8. Does the script include functionality for adjusting the parameters of the Chroma database based on session performance metrics?

References

- *Crowdsourcing Multiple Choice Science Questions* by Johannes Welbl, Nelson F. Liu, Matt Gardner: http://arxiv.org/abs/1707.06209.
- *MiniLMv2: Multi-Head Self-Attention Relation Distillation for Compressing Pretrained Transformers* by Wenhui Wang, Hangbo Bao, Shaohan Huang, Li Dong, Furu Wei: https://arxiv.org/abs/2012.15828.
- Hugging Face Llama model documentation: https://huggingface.co/docs/transformers/main/en/model_doc/llama.
- ONNX: https://onnxruntime.ai/.

Further reading

- *MiniLM: Deep Self-Attention Distillation for Task-Agnostic Compression of Pre-Trained Transformers* by Wenhui Wang, Furu Wei, Li Dong, Hangbo Bao, Nan Yang, Ming Zhou: `https://arxiv.org/abs/2002.10957`.

- *LLaMA: Open and Efficient Foundation Language Models* by Hugo Touvron, Thibaut Lavril, Gautier Lzacard, et al.: `https://arxiv.org/abs/2302.13971`.

- Building an ONNX Runtime package: `https://onnxruntime.ai/docs/build/custom.html#custom-build-packages`.

Unlock this book's exclusive benefits now

This book comes with additional benefits designed to elevate your learning experience.

Note: Have your purchase invoice ready before you begin. `https://www.packtpub.com/unlock/9781836200918`

9

Empowering AI Models: Fine-Tuning RAG Data and Human Feedback

An organization that continually increases the volume of its RAG data will reach the threshold of non-parametric data (not pretrained on an LLM). At that point, the mass of RAG data accumulated might become extremely challenging to manage, posing issues related to storage costs, retrieval resources, and the capacity of the generative AI models themselves. Moreover, a pretrained generative AI model is trained up to a cutoff date. The model ignores new knowledge starting the very next day. This means that it will be impossible for a user to interact with a chat model on the content of a newspaper edition published after the cutoff date. That is when retrieval has a key role to play in providing RAG-driven content.

Companies like Google, Microsoft, Amazon, and other web giants may require exponential data and resources. Certain domains, such as the legal rulings in the United States, may indeed require vast amounts of data. However, this doesn't apply to a wide range of domains. Many corporations do not need to maintain such large datasets, and in some cases, large portions of static data—like those in hard sciences—can remain stable for a long time. Such static data can be fine-tuned to reduce the volume of RAG data required.

In this chapter, therefore, we will first examine the architecture of RAG data reduction through fine-tuning. We will focus on a dataset that contains ready-to-use documents but also stresses the human-feedback factor. We will demonstrate how to transform non-parametric data into parametric, fine-tuned data in an OpenAI model. Then, we will download and prepare the dataset from the previous chapter, converting the data into well-formatted prompt and completion pairs for fine-tuning in JSONL. We will fine-tune a cost-effective OpenAI model, GPT-4o-mini, which will prove sufficient for the completion task we will implement. Once the model is fine-tuned, we will test it on our dataset to verify that it has successfully taken our data into account. Finally, we will explore OpenAI's metrics interface, which enables us to monitor our technical metrics, such as accuracy and usage metrics, to assess the cost-effectiveness of our approach.

To sum up, this chapter covers the following topics:

- The limits of managing RAG data
- The challenge of determining what data to fine-tune
- Preparing a JSON dataset for fine-tuning
- Running OpenAI's processing tool to produce a JSONL dataset
- Fine-tuning an OpenAI model
- Managing the fine-tuning processing time
- Running the fine-tuned model

Let's begin by defining the architecture of the fine-tuning process.

The architecture of fine-tuning static RAG data

In this section, we question the usage of non-parametric RAG data when it exceeds a manageable threshold, as described in the *RAG versus fine-tuning* section in *Chapter 1, Why Retrieval Augmented Generation?*, which stated the principle of a threshold. *Figure 9.1* adapts the principle to this section:

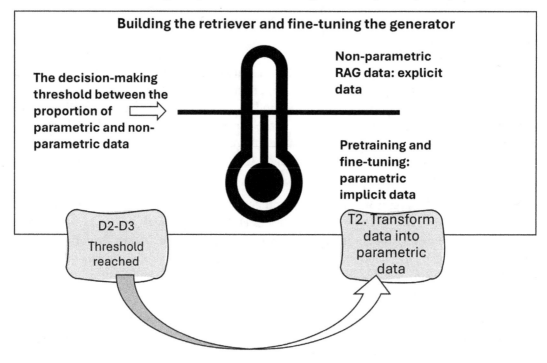

Figure 9.1: Fine-tuning threshold reached for RAG data

Notice that the processing (**D2**) and storage (**D3**) thresholds have been reached for static data versus the dynamic data in the RAG data environment. The threshold depends on each project and parameters such as:

- **The volume of RAG data to process**: Embedding data requires human and machine resources. Even if we don't embed the data, piling up static data (data that is stable over a long period of time) makes no sense.

- **The volume of RAG data to store and retrieve**: At some point, if we keep stacking data up, much of it may overlap.

- **The retrievals require resources**: Even if the system is open source, there is still an increasing number of resources to manage.

Other factors, too, may come into play for each project. Whatever the reason, fine-tuning can be a good solution when we reach the RAG data threshold.

The RAG ecosystem

In this section, we will return to the RAG ecosystem described in *Chapter 1*. We will focus on the specific components we need for this chapter. The following figure presents the fine-tuning components in color and the ones we will not need in gray:

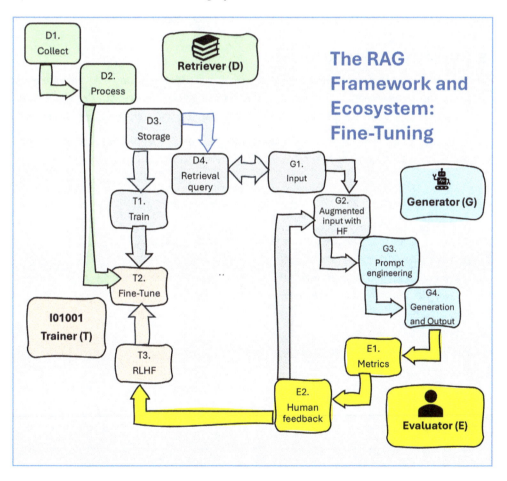

Figure 9.2: Fine-tuning components of the RAG ecosystem

The key features of the fine-tuning ecosystems we will build can be summarized in the following points:

- **Collecting (D1) and preparing (D2) the dataset:** We will download and process the human-crafted crowdsourced SciQ hard science dataset we implemented in the previous chapter: https://huggingface.co/datasets/sciq.

- **Human feedback (E2):** We can assume that human feedback played an important role in the SciQ hard science dataset. The dataset was controlled by humans and updated so we can think of it as a simulation of how reliable human feedback can be fine-tuned to alleviate the volume of RAG datasets. We can go further and say it is possible that, in real-life projects, the explanations present in the SciQ dataset can sometimes come from human evaluations of models, as we explored in *Chapter 5, Boosting RAG Performance with Expert Human Feedback*.

- **Fine-tuning (T2):** We will fine-tune a cost-effective OpenAI model, GPT-4o-mini.

- **Prompt engineering (G3) and generation and output (G4):** We will engineer the prompts as recommended by OpenAI and display the output.

- **Metrics (E1):** We will look at the main features of OpenAI's Metrics interface.

Let's now go to our keyboards to collect and process the SciQ dataset.

Installing the environment

Installing an environment has become complex with the rapid evolution of AI and cross-platform dependency conflicts, as we saw in *Chapter 2, RAG Embedding Vector Stores with Deep Lake and OpenAI*, in the *Setting up the environment* section. We will thus freeze the package versions when possible.

For this program, open the Fine_tuning_OpenAI_GPT_4o_mini.ipynb notebook in the Chapter09 directory on GitHub. The program first retrieves the OpenAI API key:

```
#You can retrieve your API key from a file(1)
# or enter it manually(2)


#Comment this cell if you want to enter your key manually.
#(1)Retrieve the API Key from a file
#Store you key in a file and read it(you can type it directly in the notebook
but it will be visible for somebody next to you)
from google.colab import drive
drive.mount('/content/drive')
f = open("drive/MyDrive/files/api_key.txt", "r")
API_KEY=f.readline()
f.close()
```

We then install openai and set the API key:

```
try:
    import openai
except:
```

```
  !pip install openai==1.42.0
  import openai
#(2) Enter your manually by
# replacing API_KEY by your key.
#The OpenAI Key
import os
os.environ['OPENAI_API_KEY'] =API_KEY
openai.api_key = os.getenv("OPENAI_API_KEY")
```

Now, we install `jsonlines` to generate JSONL data:

```
  !pip install jsonlines==4.0.0
```

We now install `datasets`:

```
  !pip install datasets==2.20.0
```

Read the *Installing the environment* section of *Chapter 8, Dynamic RAG with Chroma and Hugging Face Llama,* for explanations of the dependency conflicts involved when installing `datasets`.

Some issues with the installation may occur but the dataset will be downloaded anyway. We must expect and accept such issues as the leading platforms continually update their packages and create conflicts with pre-installed environments such as Google Colab. You can create a special environment for this program. Bear in mind that your other programs might encounter issues due to other package constraints.

We are now ready to prepare the dataset.

1. Preparing the dataset for fine-tuning

Fine-tuning an OpenAI model requires careful preparation; otherwise, the fine-tuning job will fail. In this section, we will carry out the following steps:

1. Download the dataset from Hugging Face and prepare it by processing its columns.
2. Stream the dataset to a JSON file in JSONL format.

The program begins by downloading the dataset.

1.1. Downloading and visualizing the dataset

We will download the SciQ dataset we embedded in *Chapter 8*. As we saw, embedding thousands of documents takes time and resources. In this section, we will download the dataset, but this time, *we will not embed it.* We will let the OpenAI model handle that for us while fine-tuning the data.

The program downloads the same Hugging Face dataset as in *Chapter 8* and filters the training portion of the dataset to include only non-empty records with the correct answer and support text to explain the answer to the questions:

```
# Import required libraries
from datasets import load_dataset
```

```python
import pandas as pd

# Load the SciQ dataset from HuggingFace
dataset_view = load_dataset("sciq", split="train")

# Filter the dataset to include only questions with support and correct answer
filtered_dataset = dataset_view.filter(lambda x: x["support"] != "" and
x["correct_answer"] != "")
# Print the number of questions with support
print("Number of questions with support: ", len(filtered_dataset))
```

The preceding code then prints the number of filtered questions with support text. The output shows that we have a subset of 10,481 records:

```
Number of questions with support:  10481
```

Now, we will load the dataset to a DataFrame and drop the distractor columns (those with wrong answers to the questions):

```python
# Convert the filtered dataset to a pandas DataFrame
df_view = pd.DataFrame(filtered_dataset)

# Columns to drop
columns_to_drop = ['distractor3', 'distractor1', 'distractor2']

# Dropping the columns from the DataFrame
df_view = df.drop(columns=columns_to_drop)

# Display the DataFrame
df_view.head()
```

The output displays the three columns we need:

	question	correct_answer	support
0	What type of organism is commonly used in prep...	mesophilic organisms	Mesophiles grow best in moderate temperature, ...
1	What phenomenon makes global winds blow northe...	coriolis effect	Without Coriolis Effect the global winds would...
2	Changes from a less-ordered state to a more-or...	exothermic	Summary Changes of state are examples of phase...
3	What is the least dangerous radioactive decay?	alpha decay	All radioactive decay is dangerous to living t...
4	Kilauea in hawaii is the world's most continuo...	smoke and ash	Example 3.5 Calculating Projectile Motion: Hot...

Figure 9.3: Output displaying three columns

We need the question that will become the prompt. The correct_answer and support columns will be used for the completion. Now that we have examined the dataset, we can stream the dataset directly to a JSON file.

1.2. Preparing the dataset for fine-tuning

To train the completion model we will use, we need to write a JSON file in the very precise JSONL format as required.

We download and process the dataset in the same way as we did to visualize it in the *1.1. Downloading and visualizing the dataset* section, which is recommended to check the dataset before fine-tuning it.

We now write the messages for GPT-4o-mini in JSONL:

```
# Prepare the data items for JSON lines file
items = []
for idx, row in df.iterrows():
    detailed_answer = row['correct_answer'] + " Explanation: " + row['support']
    items.append({
        "messages": [
            {"role": "system", "content": "Given a science question, provide
the correct answer with a detailed explanation."},
            {"role": "user", "content": row['question']},
            {"role": "assistant", "content": detailed_answer}
        ]
    })
```

We first define the detailed answer (detailed_answer) with the correct answer ('correct_answer') and a supporting (support) explanation.

Then we define the messages (messages) for the GPT-4o-mini model:

- {"role": "system", "content": ...}: This sets the initial instruction for the language model, telling it to provide detailed answers to science questions.
- {"role": "user", "content": row['question']}: This represents the user asking a question, taken from the question column of the DataFrame.
- {"role": "assistant", "content": detailed_answer}: This represents the assistant's response, providing the detailed answer constructed earlier.

We can now write our JSONL dataset to a file:

```
# Write to JSON lines file
with jsonlines.open('/content/QA_prompts_and_completions.json', 'w') as writer:
    writer.write_all(items)
```

We have given the OpenAI model a structure it expects and has been trained to understand. We can load the JSON file we just created in a pandas DataFrame to verify its content:

```
dfile="/content/QA_prompts_and_completions.json"
import pandas as pd
# Load the data
```

```
df = pd.read_json(dfile, lines=True)
df
```

The following excerpt of the file shows that we have successfully prepared the JSON file:

index	messages
0	{'role': 'system', 'content': 'Given a science question, provide the correct answer with a detailed explanation.'},{'role': 'user', 'content': 'What type of organism is commonly used in preparation of foods such as cheese and yogurt?'}, {'role': 'assistant', 'content': 'mesophilic organisms Explanation: Mesophiles grow best in moderate temperature, typically between 25°C and 40°C (77°F and 104°F). Mesophiles are often found living in or on the bodies of humans or other animals. The optimal growth temperature of many pathogenic mesophiles is 37°C (98°F), the normal human body temperature. Mesophilic organisms have important uses in food preparation, including cheese, yogurt, beer and wine.'}
1	{'role': 'system', 'content': 'Given a science question, provide the correct answer with a detailed explanation.'},{'role': 'user', 'content': 'What phenomenon makes global winds blow northeast to southwest or the reverse in the northern hemisphere and northwest to southeast or the reverse in the southern hemisphere?'},{'role': 'assistant', 'content': 'coriolis effect Explanation: Without Coriolis Effect the global winds would blow north to south or south to north. But Coriolis makes them blow northeast to southwest or the reverse in the Northern Hemisphere. The winds blow northwest to southeast or the reverse in the southern hemisphere.'}

Figure 9.4: File excerpt

That's it! We are now ready to run a fine-tuning job.

2. Fine-tuning the model

To train the model, we retrieve our training file and create a fine-tuning job. We begin by creating an OpenAI client:

```
from openai import OpenAI
import jsonlines
client = OpenAI()
```

Then we use the file we generated to create another training file that is uploaded to OpenAI:

```
# Uploading the training file

result_file = client.files.create(
    file=open("QA_prompts_and_completions.json", "rb"),
    purpose="fine-tune"
)
```

We print the file information for the dataset we are going to use for fine-tuning:

```
print(result_file)
param_training_file_name = result_file.id
print(param_training_file_name)
```

We now create and display the fine-tuning job:

```
# Creating the fine-tuning job
```

```
ft_job = client.fine_tuning.jobs.create(
  training_file=param_training_file_name,
  model="gpt-4o-mini-2024-07-18"
)


# Printing the fine-tuning job
print(ft_job)
```

The output first provides the name of the file, its purpose, its status, and the OpenAI name of the file ID:

```
FileObject(id='file-EUPGmm1yAd3axrQ0pyoeAKuE', bytes=8062970, created_
at=1725289249, filename='QA_prompts_and_completions.json', object='file',
purpose='fine-tune', status='processed', status_details=None) file-
EUPGmm1yAd3axrQ0pyoeAKuE
```

The code displays the details of the fine-tuning job:

```
FineTuningJob(id='ftjob-O1OEE7eEyFNJsO2Eu5otzWA8', created_at=1725289250,
error=Error(code=None, message=None, param=None), fine_tuned_model=None,
finished_at=None, hyperparameters=Hyperparameters(n_epochs='auto', batch_
size='auto', learning_rate_multiplier='auto'), model='gpt-4o-mini-2024-07-18',
object='fine_tuning.job', organization_id='org-h2Kjmcir4wyGtqq1mJALLGIb',
result_files=[], seed=1103096818, status='validating_files', trained_
tokens=None, training_file='file-EUPGmm1yAd3axrQ0pyoeAKuE', validation_
file=None, estimated_finish=None, integrations=[], user_provided_suffix=None)
```

The output provides the details we need to monitor the job. Here is a brief description of some of the key-value pairs in the output:

- `Job ID`: ftjob-O1OEE7eEyFNJsO2Eu5otzWA8.
- `Status`: validating_files. This means OpenAI is currently checking the training file to make sure it's suitable for fine-tuning.
- `Model`: gpt-4o-mini-2024-07-18. We're using a smaller, more cost-effective version of GPT-4 for fine-tuning.
- `Training File`: file-EUPGmm1yAd3axrQ0pyoeAKuE. This is the file we've provided that contains the examples to teach the model.

Some key hyperparameters are:

- `n_epochs`: 'auto': OpenAI will automatically determine the best number of training cycles.
- `batch_size`: 'auto': OpenAI will automatically choose the optimal batch size for training.
- `learning_rate_multiplier`: 'auto': OpenAI will automatically adjust the learning rate during training.
- `Created at`: 2024-06-30 08:20:50.

This information will prove useful if you wish to perform an in-depth study of fine-tuning OpenAI models. We can also use it to monitor and manage our fine-tuning process.

2.1. Monitoring the fine-tunes

In this section, we will extract the minimum information we need to monitor the jobs for all our fine-tunes. We will first query OpenAI to obtain the three latest fine-tuning jobs:

```python
import pandas as pd
from openai import OpenAI
client = OpenAI()
# Assume client is already set up and authenticated
response = client.fine_tuning.jobs.list(limit=3) # increase to include your
history
```

We then initialize the lists of information we want to visualize:

```python
# Initialize lists to store the extracted data
job_ids = []
created_ats = []
statuses = []
models = []
training_files = []
error_messages = []
fine_tuned_models = [] # List to store the fine-tuned model names
```

Following that, we iterate through response to retrieve the information we need:

```python
# Iterate over the jobs in the response
for job in response.data:
    job_ids.append(job.id)
    created_ats.append(job.created_at)
    statuses.append(job.status)
    models.append(job.model)
    training_files.append(job.training_file)
    error_message = job.error.message if job.error else None
    error_messages.append(error_message)

    # Append the fine-tuned model name
    fine_tuned_model = job.fine_tuned_model if hasattr(job, 'fine_tuned_model')
    else None
    fine_tuned_models.append(fine_tuned_model)
```

We now create a DataFrame with the information we extracted:

```python
import pandas as pd
# Assume client is already set up and authenticated
response = client.fine_tuning.jobs.list(limit=3)
```

```
# Create a DataFrame
df = pd.DataFrame({
    'Job ID': job_ids,
    'Created At': created_ats,
    'Status': statuses,
    'Model': models,
    'Training File': training_files,
    'Error Message': error_messages,
    'Fine-Tuned Model': fine_tuned_models # Include the fine-tuned model names
})
```

Finally, we convert the timestamps to readable format and display the list of fine-tunes and their status:

```
# Convert timestamps to readable format
df['Created At'] = pd.to_datetime(df['Created At'], unit='s')
df = df.sort_values(by='Created At', ascending=False)


# Display the DataFrame
df
```

The output provides a monitoring dashboard of the list of our jobs, as shown in *Figure 9.5*:

	Job ID	Created At	Status	Model	Training File	Error Message	Fine-Tuned Model
0	ftjob-O1OEE7eEyFNJsO2Eu5otzWA8	2024-09-02 15:00:50	running	gpt-4o-mini-2024-07-18	file-EUPGmm1yAd3axrQ0pyoeAKuE	None	None
1	ftjob-gQGiuvPMvSop0tzGaDn1NMql	2024-09-02 14:26:35	succeeded	gpt-4o-mini-2024-07-18	file-1OyEhi0D2b1kcL54JbQ3P1Pa	None	ft:gpt-4o-mini-2024-07-18:personal::A32qtJOo
2	ftjob-oVB0RAcwn3NEi4u0qOMeMZUF	2024-09-02 14:11:28	succeeded	gpt-4o-mini-2024-07-18	file-0dxQmL84uLME7ehnGljQAxit	None	ft:gpt-4o-mini-2024-07-18:personal::A32VfYlz

Figure 9.5: Job list in the pandas DataFrame

You can see that for job 0, the status of the task is running. The status informs you of the different steps of the process such as validating the files, running, failed, or succeeded. In this case, the fine-tuning process is running. If you refresh this cell regularly, you will see the status.

We will now retrieve the most recent model trained for the Fine-Tuned Model column. If the training fails, this column will be empty. If not, we can retrieve it:

```
import pandas as pd


generation=False  # until the current model is fine-tuned
# Attempt to find the first non-empty Fine-Tuned Model
```

```python
non_empty_models = df[df['Fine-Tuned Model'].notna() & (df['Fine-Tuned Model']
!= '')]

if not non_empty_models.empty:
    first_non_empty_model = non_empty_models['Fine-Tuned Model'].iloc[0]
    print("The latest fine-tuned model is:", first_non_empty_model)
    generation=True
else:
    first_non_empty_model='None'
    print("No fine-tuned models found.")
# Display the first non-empty Fine-Tuned Model in the DataFrame
first_non_empty_model = df[df['Fine-Tuned Model'].notna() & (df['Fine-Tuned
Model'] != '')]['Fine-Tuned Model'].iloc[0]
print("The lastest fine-tuned model is:", first_non_empty_model)
```

The output will display the name of the latest fine-tuned model if there is one or inform us that no fine-tuned model is found. In this case, GPT-4o-mini was successfully trained:

```
The latest fine-tuned model is: ft:gpt-4o-mini-2024-07-18:personal::A32VfYIz
```

If a fine-tuned model is found, generation=True, it will trigger the OpenAI completion calls in the following cells. If no model is found, generation=False, it will not run the OpenAI API in the rest of the notebook to avoid using models that you are not training. You can set generation to True in a new cell and then select any fine-tuned model you wish.

We know that the training job can take a while. You can refresh the pandas DataFrame from time to time. You can write code that checks the status of another job and waits for a name to appear for your training job or an error message. You can also wait for OpenAI to send you an email informing you that the training job is finished. If the training job fails, we must verify our training data for any inconsistencies, missing values, or incorrect labels. Additionally, ensure that the JSON file format adheres to OpenAI's specified schema, including correct field names, data types, and structure.

Once the training job is finished, we can run completion tasks.

3. Using the fine-tuned OpenAI model

We are now ready to use our fine-tuned OpenAI GPT-4o-mini model. We will begin by defining a prompt based on a question taken from our initial dataset:

```python
# Define the prompt
prompt = "What phenomenon makes global winds blow northeast to southwest or the
reverse in the northern hemisphere and northwest to southeast or the reverse in
the southern hemisphere?"
```

The goal is to verify whether the dataset has been properly trained and will produce results similar to the completions we defined. We can now run the fine-tuned model:

```
# Assume first_non_empty_model is defined above this snippet
if generation==True:
    response = client.chat.completions.create(
        model=first_non_empty_model,
        temperature=0.0,  # Adjust as needed for variability
        messages=[
            {"role": "system", "content": "Given a question, reply with a
complete explanation for students."},
            {"role": "user", "content": prompt}
        ]
    )
else:
    print("Error: Model is None, cannot proceed with the API request.")
```

The parameters of the request must fit our scenario:

- `model=first_non_empty_model` is our pretrained model.
- `prompt=prompt` is our predefined prompt.
- `temperature=0.0` is set to a low value because we do not want any "creativity" for this hard science completion task.

Once we run the request, we can format and display the response. The following code contains two cells to display and extract the response.

First, we can print the raw response:

```
if generation==True:
  print(response)
```

The output contains the response and information on the process:

```
ChatCompletion(id='chatcmpl-A32pvH9wLvNsSRmB1sUjxOW4Z6Xr6',…
```

We then extract the text of the response:

```
if (generation==True):
  # Access the response from the first choice
  response_text = response.choices[0].message.content
  # Print the response
  print(response_text)
```

The output is a string:

```
Coriolis effect Explanation: The Coriolis effect is…
```

Finally, we can format the response string into a nice paragraph with the Python wrapper:

```
import textwrap
```

```
if generation==True:
wrapped_text = textwrap.fill(response_text.strip(), 60)
print(wrapped_text)
```

The output shows that our data has been taken into account:

```
Coriolis effect Explanation: The Coriolis effect is a
phenomenon that causes moving objects, such as air and
water, to turn and twist in response to the rotation of the
Earth. It is responsible for the rotation of large weather
systems, such as hurricanes, and the direction of trade
winds and ocean currents. In the Northern Hemisphere, the
effect causes moving objects to turn to the right, while in
the Southern Hemisphere, objects turn to the left. The
Coriolis effect is proportional to the speed of the moving
object and the strength of the Earth's rotation, and it is
negligible for small-scale movements, such as water flowing
in a sink.
```

Let's look at the initial completion for our prompt:

coriolis effect	Without Coriolis Effect the global winds would blow north to south or south to north. But Coriolis makes them blow northeast to southwest or the reverse in the Northern Hemisphere. The winds blow northwest to southeast or the reverse in the southern hemisphere.

Figure 9.6: Initial completion

The response is thus satisfactory. This might not always be the case and might require more work on the datasets (better data, large volumes of data, etc.) incrementally until you have reached a satisfactory goal.

You can save the name of your model in a text file or anywhere you wish. You can now run your model in another program using the name of your trained model, or you can reload this notebook at any time:

1. Run the `Installing the environment` section of this notebook.
2. Define a prompt of your choice related to the dataset we trained.
3. Enter the name of your model in the OpenAI completion request.
4. Run the request and analyze the response.

You can consult OpenAI's fine-tuning documentation for further information if necessary: `https://platform.openai.com/docs/guides/fine-tuning/fine-tuning`.

Metrics

OpenAI provides a user interface to analyze the metrics of the training process and model. You can access the metrics related to your fine-tuned models at `https://platform.openai.com/finetune/`.

The interface displays the list of your fine-tuned jobs:

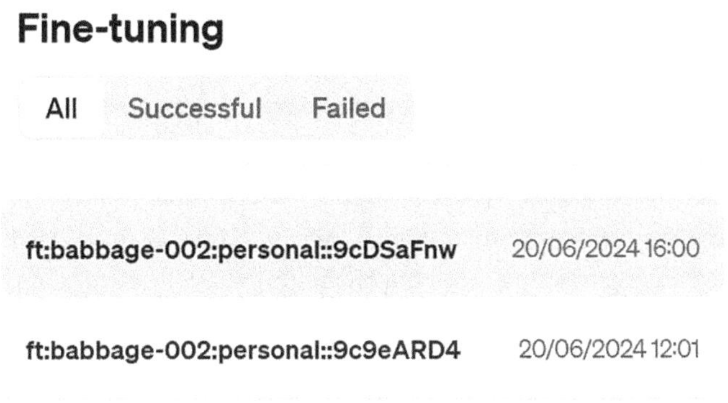

Fine-tuning

All   Successful   Failed

ft:babbage-002:personal::9cDSaFnw   20/06/2024 16:00

ft:babbage-002:personal::9c9eARD4   20/06/2024 12:01

Figure 9.7: List of fine-tuned jobs

You can choose to view all the fine-tuning jobs, the ones that were successful, or the ones that failed. If we choose a job that was successful, for example, we can view the job details as shown in the following excerpt:

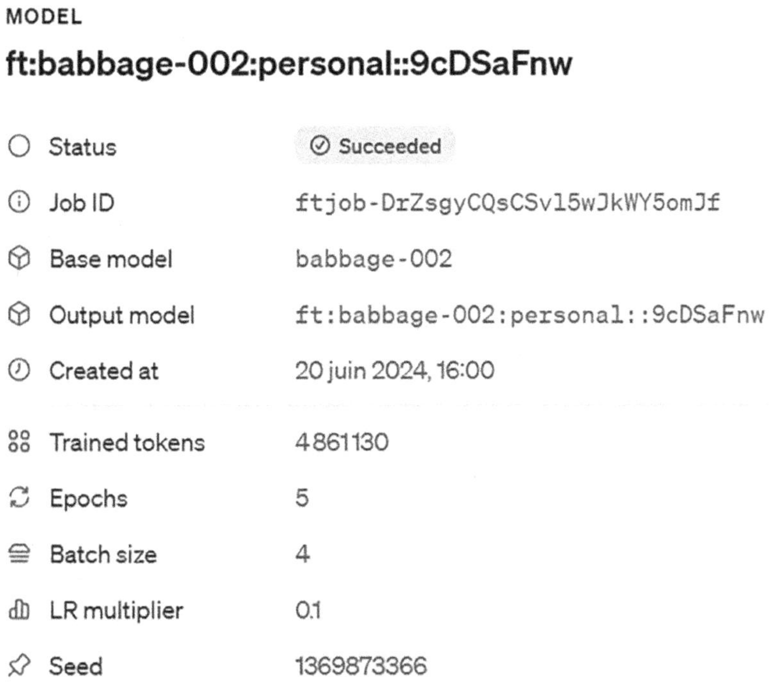

MODEL

ft:babbage-002:personal::9cDSaFnw

○	Status	⊘ Succeeded
ⓘ	Job ID	`ftjob-DrZsgyCQsCSvl5wJkWY5omJf`
⬡	Base model	`babbage-002`
⬡	Output model	`ft:babbage-002:personal::9cDSaFnw`
⦿	Created at	20 juin 2024, 16:00
⣿	Trained tokens	4 861 130
↻	Epochs	5
⬓	Batch size	4
⏸	LR multiplier	0.1
⬧	Seed	1369873366

Figure 9.8: Example view

Let's go through the information provided in this figure:

- **Status:** Indicates the status of the fine-tuning process. In this case, we can see that the process was completed successfully.
- **Job ID:** A unique identifier for the fine-tuning job. This can be used to reference the job in queries or for support purposes.
- **Base model:** Specifies the pretrained model used as the starting point for fine-tuning. In this case, gpt-4o-mini is a version of OpenAI's models.
- **Output model:** This is the identifier for the model resulting from the fine-tuning. It incorporates changes and optimizations based on the specific training data provided.
- **Created at:** The date and time when the fine-tuning job was initiated.
- **Trained tokens:** The total number of tokens (pieces of text, such as words or punctuation) that were processed during training. This metric helps gauge the extent of training.
- **Epochs:** The number of complete passes the training data went through during fine-tuning. More epochs can lead to better learning but too many may lead to overfitting.
- **Batch size:** The number of training examples utilized in one iteration of model training. Smaller batch sizes can offer more updates and refined learning but may take longer to train.
- **LR multiplier:** This refers to the learning rate multiplier, affecting how much the learning rate for the base model is adjusted during the fine-tuning process. A smaller multiplier can lead to smaller, more conservative updates to model weights.
- **Seed:** A seed for the random number generator used in the training process. Providing a seed ensures that the training process is reproducible, meaning you can get the same results with the same input conditions.

This information will help tailor the fine-tuning jobs to meet the specific needs of a project and explore alternative approaches to optimization and customization. In addition, the interface contains more information that we can explore to get an in-depth vision of the fine-tuning process. If we scroll down on the **Information** tab of our model, we can see metrics as shown here:

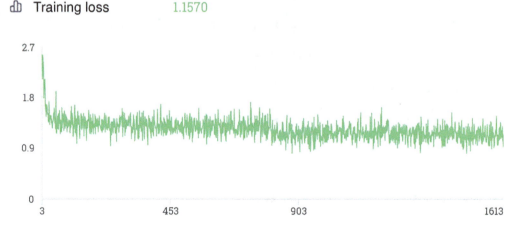

Figure 9.9: Metrics for a fine-tuned model

Training loss and the other available information can guide our training strategies (data, files, and parameters).

Training loss is a reliable metric used to evaluate the performance of a machine learning model during training. In this case, `Training loss (1.1570)` represents the model's average error on the training dataset. Lower training loss values indicate that the model is better fitting the training data. A training loss of `1.1570` suggests that the model has learned to predict or classify its training data well during the fine-tuning process.

We can also examine these values with the `Time` and `Step` information:

Messages	Metrics	
Time	**Step**	**Training loss**
18:22:58	1613	1.1570
18:22:55	1612	0.9320
18:22:39	1611	1.0838
18:22:37	1610	1.0339
18:22:37	1609	1.2993
18:22:37	1608	1.1136
18:22:35	1607	1.1126
18:22:35	1606	1.2236
18:22:34	1605	1.0821

Figure 9.10: Training loss during the training job

We must also measure the usage to monitor the cost per period and model. OpenAI provides a detailed interface at `https://platform.openai.com/usage`.

Fine-tuning can indeed be an effective way to optimize RAG data if we make sure to train a model with high-quality data and the right parameters. Now, it's time for us to summarize our journey and move to our next RAG-driven generative AI implementation.

Summary

This chapter's goal was to show that as we accumulate RAG data, some data is dynamic and requires constant updates, and as such, cannot be fine-tuned easily. However, some data is static, meaning that it will remain stable for long periods of time. This data can become parametric (stored in the weights of a trained LLM).

We first downloaded and processed the SciQ dataset, which contains hard science questions. This stable data perfectly suits fine-tuning. It contains a question, answer, and support (explanation) structure, which makes the data effective for fine-tuning. Also, we can assume human feedback was required. We can even go as far as imagining this feedback could be provided by analyzing generative AI model outputs.

We converted the data we prepared into prompts and completions in a JSONL file following the recommendations of OpenAI's preparation tool. The structure of JSONL was meant to be compatible with a completion model (prompt and completion) such as `GPT-4o-mini`. The program then fine-tuned the cost-effective `GPT-4o-mini` OpenAI model, following which we ran the model and found that the output was satisfactory. Finally, we explored the metrics of the fine-tuned model in the OpenAI metrics user interface.

We can conclude that fine-tuning can optimize RAG data in certain cases when necessary. However, we will take this process further in the next chapter, *Chapter 10, RAG for Video Stock Production with Pinecone and OpenAI*, when we run the full-blown RAG-driven generative AI ecosystem.

Questions

Answer the following questions with yes or no:

1. Do all organizations need to manage large volumes of RAG data?
2. Is the GPT-4o-mini model described as insufficient for fine-tuning tasks?
3. Can pretrained models update their knowledge base after the cutoff date without retrieval systems?
4. Is it the case that static data never changes and thus never requires updates?
5. Is downloading data from Hugging Face the only source for preparing datasets?
6. Is all RAG data eventually embedded into the trained model's parameters according to the document?
7. Does the chapter recommend using only new data for fine-tuning AI models?
8. Is the OpenAI Metrics interface used to adjust the learning rate during model training?
9. Can the fine-tuning process be effectively monitored using the OpenAI dashboard?
10. Is human feedback deemed unnecessary in the preparation of hard science datasets such as SciQ?

References

*   OpenAI fine-tuning documentation: `https://platform.openai.com/docs/guides/fine-tuning/`
*   OpenAI pricing: `https://openai.com/api/pricing/`

Further reading

- *Test of Fine-Tuning GPT by Astrophysical Data* by Yu Wang et al. is an interesting article on fine-tuning hard science data, which requires careful data preparation: `https://arxiv.org/pdf/2404.10019`

Learn further in a live workshop with the author

If you are curious how RAG-based systems scale into agentic workflows using context engineering, consider registering for the author's live workshop on 24 January 2026, which extends the ideas discussed here into hands-on practice.

`https://packt.link/gk5Yu`

10

RAG for Video Stock Production with Pinecone and OpenAI

Human creativity goes beyond the range of well-known patterns due to our unique ability to break habits and invent new ways of doing anything, anywhere. Conversely, Generative AI relies on our well-known established patterns across an increasing number of fields without really "creating" but rather replicating our habits. In this chapter, therefore, when we use the term "create" as a practical term, we only mean "generate." Generative AI, with its efficiency in automating tasks, will continue its expansion until it finds ways of replicating any human task it can. We must, therefore, learn how these automated systems work to use them for the best in our projects. Think of this chapter as a journey into the architecture of RAG in the cutting-edge hybrid human and AI agent era we are living in. We will assume the role of a start-up aiming to build an AI-driven downloadable stock of online videos. To achieve this, we will establish a team of AI agents that will work together to create a stock of commented and labeled videos.

Our journey begins with the Generator agent in *Pipeline 1: The Generator and the Commentator*. The Generator agent creates world simulations using Sora, an OpenAI text-to-video model. You'll see how the *inVideo* AI application, powered by Sora, engages in "ideation," transforming an idea into a video. The Commentator agent then splits the AI-generated videos into frames and generates technical comments with an OpenAI vision model. Next, in *Pipeline 2: The Vector Store Administrator,* we will continue our journey and build the Vector Store Administrator that manages Pinecone. The Vector Store Administrator will embed the technical video comments generated by the Commentator, upsert the vectorized comments, and query the Pinecone vector store to verify that the system is functional. Finally, we will build the Video Expert that processes user inputs, queries the vector store, and retrieves the relevant video frames. Finally, in *Pipeline 3: The Video Expert*, the Video Expert agent will augment user inputs with the raw output of the query and activate its expert OpenAI GPT-4o model, which will analyze the comment, detect imperfections, reformulate it more efficiently, and provide a label for the video.

By the end of the chapter, you will know how to automatically generate a stock of short videos by automating the process of going from raw footage to videos with descriptions and labels. You'll be able to offer a service where users can simply type a few words and obtain a video with a custom, real-time description and label.

Summing that up, this chapter covers the following topics:

- Designing Generative AI videos and comments
- Splitting videos into frames for OpenAI's vision analysis models
- Embedding the videos and upserting the vectors to a Pinecone index
- Querying the vector store
- Improving and correcting the video comments with OpenAI GPT-4o
- Automatically labeling raw videos
- Displaying the full result of the raw video process with a commented and labeled video
- Evaluating outputs and implementing metric calculations

Let's begin by defining the architecture of RAG for video production.

The architecture of RAG for video production

Automating the process of real-world video generation, commenting, and labeling is extremely relevant in various industries, such as media, marketing, entertainment, and education. Businesses and creators are continuously seeking efficient ways to produce and manage content that can scale with growing demand. In this chapter, you will acquire practical skills that can be directly applied to meet these needs.

The goal of our RAG video production use case in this chapter is to process AI-generated videos using AI agents to create a video stock of labeled videos to identify them. The system will also dynamically generate custom descriptions by pinpointing AI-generated technical comments on specific frames within the videos that fit the user input. *Figure 10.1* illustrates the AI-agent team that processes RAG for video production:

RAG for Video Production
**From raw video files to dynamically generated
descriptions and automated labeling**

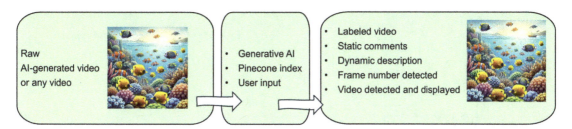

Figure 10.1: From raw videos to labeled and commented videos

We will implement AI agents for our RAG video production pipeline that will:

- Generate raw videos automatically and download them
- Split the videos into frames
- Analyze a sample of frames
- Activate an OpenAI LLM model to generate technical comments
- Save the technical comments with a unique index, the comment itself, the frame number analyzed, and the video file name
- Upsert the data in a Pinecone index vector store
- Query the Pinecone vector store with user inputs
- Retrieve the specific frame within a video that is most similar to its technical comment
- Augment the user input with the technical comment of the retrieved frame
- Ask the OpenAI LLM to analyze the logic of the technical comment that may contain contradictions and imperfections detected in the video and then produce a dynamic, well-tailored description of the video with the frame number and the video file name
- Display the selected video
- Evaluate the outputs and apply metric calculations

We will thus go from raw videos to labeled videos with tailored descriptions based on the user input. For example, we will be able to ask precise questions such as the following:

```
"Find a basketball player that is scoring with a dunk."
```

This means that the system will be able to find a frame (image) within the initially unlabeled video, select the video, display it, and generate a tailored comment dynamically. To attain our goal, we will implement AI agents in three pipelines, as illustrated in the following figure:

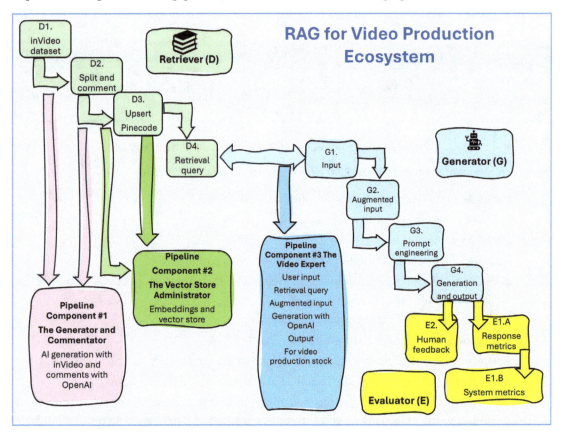

Figure 10.2: The RAG for Video Production Ecosystem with Generative AI agents

Now, what you see in the figure above is:

- **Pipeline 1:** The **Generator** and the **Commentator**

  The **Generator** produces AI-generated videos with OpenAI Sora. The **Commentator** splits the videos into frames that are commented on by one of OpenAI's vision models. The **Commentator** agent then saves the comments.

- **Pipeline 2:** The **Vector Store Administrator**

  This pipeline will embed and upsert the comments made by *Pipeline 1* to a Pinecone index.

- **Pipeline 3: The Video Expert**

  This pipeline will query the Pinecone vector store based on user input. The query will return the most similar frame within a video, augment the input with the technical comment, and ask OpenAI GPT-4o to find logic imperfections in the video, point them out, and then produce a tailored comment of the video for the user and a label. This section also contains evaluation functions (the Evaluator) and metric calculations.

 Time measurement functions are encapsulated in several of the key functions of the preceding ecosystem.

The RAG video production system we will build allows indefinite scaling by processing one video at a time, using only a CPU and little memory, while leveraging Pinecone's storage capacity. This effectively demonstrates the concept of automated video production, but implementing this production system in a real-life project requires hard work. However, the technology is there, and the future of video production is undergoing a historical evolution. Let's dive into the code, beginning with the environment.

The environment of the video production ecosystem

The Chapter10 directory on GitHub contains the environment for all four notebooks in this chapter:

- `Videos_dataset_visualization.ipynb`
- `Pipeline_1_The_Generator_and_the_Commentator.ipynb`
- `Pipeline_2_The_Vector_Store_Administrator.ipynb`
- `Pipeline_3_The_Video_Expert.ipynb`

Each notebook includes an *Installing the environment* section, including a set of the following sections that are identical across all notebooks:

- *Importing modules and libraries*
- *GitHub*
- *Video download and display functions*
- *OpenAI*
- *Pinecone*

This chapter aims to establish a common pre-production installation policy that will focus on the pipelines' content once we dive into the RAG for video production code. This policy is limited to the scenario described in this chapter and will vary depending on the requirements of each real-life production environment.

The notebooks in this chapter only require a CPU, limited memory, and limited disk space. As such, the whole process can be streamlined indefinitely one video at a time in an optimized, scalable environment.

Let's begin by importing the modules and libraries we need for our project.

Importing modules and libraries

The goal is to prepare a pre-production global environment common to all the notebooks. As such, the modules and libraries are present in all four notebooks regardless of whether they are used or not in a specific program:

```
from IPython.display import HTML # to display videos
import base64 # to encode videos as base64
from base64 import b64encode # to encode videos as base64
import os # to interact with the operating system
import subprocess # to run commands
import time # to measure execution time
import csv # to save comments
import uuid # to generate unique ids
import cv2 # to split videos
from PIL import Image # to display videos
import pandas as pd # to display comments
import numpy as np # to use Numerical Python
from io import BytesIO #to manage a binary stream of data in memory
```

Each of the four notebooks contains these modules and libraries, as shown in the following table:

Code	Comment
`from IPython.display import HTML`	To display videos
`import base64`	To encode videos as `base64`
`from base64 import b64encode`	To encode videos as `base64`
`import os`	To interact with the operating system
`import subprocess`	To run commands
`import time`	To measure execution time
`import csv`	To save comments
`import uuid`	To generate unique IDs
`import cv2`	To split videos (open source computer vision library)
`from PIL import Image`	To display videos

import pandas as pd	To display comments
import numpy as np	To use Numerical Python
from io import BytesIO	For a binary stream of data in memory

Table 10.1: Modules and libraries for our video production system

The Code column contains the module or library name, while the Comment column provides a brief description of their usage. Let's move on to GitHub commands.

GitHub

download(directory, filename) is present in all four notebooks. The main function of download(directory, filename) is to download the files we need from the book's GitHub repository:

```
def download(directory, filename):
    # The base URL of the image files in the GitHub repository
    base_url = 'https://raw.githubusercontent.com/Denis2054/RAG-Driven-
Generative-AI/main/'

    # Complete URL for the file
    file_url = f"{base_url}{directory}/{filename}"

    # Use curl to download the file
    try:
        # Prepare the curl command
        curl_command = f'curl -o {filename} {file_url}'

        # Execute the curl command
        subprocess.run(curl_command, check=True, shell=True)
        print(f"Downloaded '{filename}' successfully.")
    except subprocess.CalledProcessError:
        print(f"Failed to download '{filename}'. Check the URL, your internet
connection, and if the token is correct and has appropriate permissions.")
```

The preceding function takes two arguments:

- directory, which is the GitHub directory that the file we want to download is located in
- filename, which is the name of the file we want to download

OpenAI

The OpenAI package is installed in all three pipeline notebooks but not in Video_dataset_visualization. ipynb, which doesn't require an LLM. You can retrieve the API key from a file or enter it manually (but it will be visible):

```
#You can retrieve your API key from a file(1)
```

```
# or enter it manually(2)
#Comment this cell if you want to enter your key manually.

#(1)Retrieve the API Key from a file
#Store you key in a file and read it(you can type it directly in the notebook
but it will be visible for somebody next to you)
from google.colab import drive
drive.mount('/content/drive')
f = open("drive/MyDrive/files/api_key.txt", "r")
API_KEY=f.readline()o
Nf.close()
```

You will need to sign up at www.openai.com before running the code and obtain an API key. The program installs the openai package:

```
try:
    import openai
except:
    #!pip install openai==1.45.0
    import openai
```

Finally, we set an environment variable for the API key:

```
#(2) Enter your manually by
# replacing API_KEY by your key.
#The OpenAI Key
os.environ['OPENAI_API_KEY'] =API_KEY
openai.api_key = os.getenv("OPENAI_API_KEY")
```

Pinecone

The *Pinecone* section is only present in `Pipeline_2_The_Vector_Store_Administrator.ipynb` and `Pipeline_3_The_Video_Expert.ipynb` when the Pinecone vector store is required. The following command installs Pinecone, and then Pinecone is imported:

```
!pip install pinecone-client==4.1.1
import pinecone
```

The program then retrieves the key from a file (or you can enter it manually):

```
f = open("drive/MyDrive/files/pinecone.txt", "r")
PINECONE_API_KEY=f.readline()
f.close()
```

In production, you can set an environment variable or implement the method that best fits your project so that the API key is never visible.

 The *Evaluator* section of `Pipeline_3_The_Video_Expert.ipynb` contains its own requirements and installations.

With that, we have defined the environment for all four notebooks, which contain the same sub-sections we just described in their respective *Installing the environment* sections. We can now fully focus on the processes involved in the video production programs. We will begin with the Generator and Commentator.

Pipeline 1: Generator and Commentator

A revolution is on its way in computer vision with automated video generation and analysis. We will introduce the Generator AI agent with Sora in *The AI-generated video dataset* section. We will explore how OpenAI Sora was used to generate the videos for this chapter with a text-to-video diffusion transformer. The technology itself is something we have expected and experienced to some extent in professional film-making environments. However, the novelty relies on the fact that the software has become mainstream in a few clicks, with inVideo, for example!

In the *The Generator and the Commentator* section, we will extend the scope of the Generator to collecting and processing the AI-generated videos. The Generator splits the videos into frames and works with the Commentator, an OpenAI LLM, to produce comments on samples of video frames.

The Generator's task begins by producing the AI-generated video dataset.

The AI-generated video dataset

The first AI agent in this project is a text-to-video diffusion transformer model that generates a video dataset we will implement. The videos for this chapter were specifically generated by Sora, a text-to-video AI model released by OpenAI in February 2024. You can access Sora to view public AI-generated videos and create your own at `https://ai.invideo.io/`. AI-generated videos also allow for free videos with flexible copyrig0ht terms that you can check out at `https://invideo.io/terms-and-conditions/`.

 Once you have gone through this chapter, you can also create your own video dataset with any source of videos, such as smartphones, video stocks, and social media.

AI-generated videos enhance the speed of creating video datasets. Teams do not have to spend time finding videos that fit their needs. They can obtain a video quickly with a prompt that can be an idea expressed in a few words. AI-generated videos represent a huge leap into the future of AI applications. Sora's potential applies to many industries, including filmmaking, education, and marketing. Its ability to generate nuanced video content from simple text prompts opens new avenues for creative and educational outputs.

Although AI-generated videos (and, in particular, diffusion transformers) have changed the way we create world simulations, this represents a risk for jobs in many areas, such as filmmaking. The risk of deep fakes and misinformation is real. At a personal level, we must take ethical considerations into account when we implement Generative AI in a project, thus producing constructive, ethical, and realistic content.

Let's see how a diffusion transformer can produce realistic content.

How does a diffusion transformer work?

At the core of Sora, as described by Liu et al., 2024 (see the *References* section), is a diffusion transformer model that operates between an encoder and a decoder. It uses user text input to guide the content generation, associating it with patches from the encoder. The model iteratively refines these noisy latent representations, enhancing their clarity and coherence. Finally, the refined data is passed to the decoder to reconstruct high-fidelity video frames. The technology involved includes vision transformers such as CLIP and LLMs such as GPT-4, as well as other components OpenAI continually includes in its vision model releases.

The encoder and decoder are integral components of the overall diffusion model, as illustrated in *Figure 10.3*. They both play a critical role in the workflow of the transformer diffusion model:

- **Encoder:** The encoder's primary function is to compress input data, such as images or videos, into a lower-dimensional latent space. The encoder thus transforms high-dimensional visual data into a compact representation while preserving crucial information. A lower-dimensional latent space obtained is a compressed representation of high-dimensional data, retaining essential features while reducing complexity. For example, a high-resolution image (1024x1024 pixels, 3 color channels) can be compressed by an encoder into a vector of 1000 values, capturing key details like shape and texture. This makes processing and manipulating images more efficient.
- **Decoder:** The decoder reconstructs the original data from the latent representation produced by the encoder. It performs the encoder's reverse operation, transforming the low-dimensional latent space back into high-dimensional pixel space, thus generating the final output, such as images or videos.

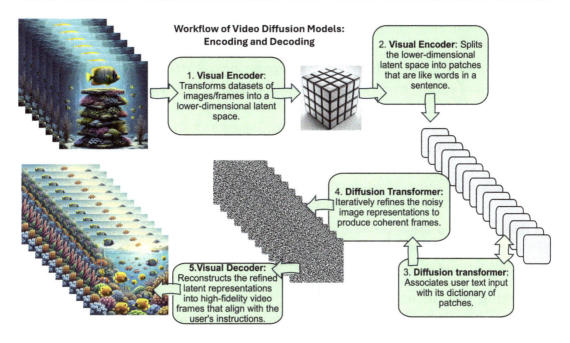

Figure 10.3: The encoding and decoding workflow of video diffusion models

💡 **Quick tip:** Enhance your coding experience with the **AI Code Explainer** and **Quick Copy** features. Open this book in the next-gen Packt Reader. Click the **Copy** button (**1**) to quickly copy code into your coding environment, or click the **Explain** button (**2**) to get the AI assistant to explain a block of code to you.

Copy   Explain

```
function calculate(a, b) {
    return {sum: a + b};
};
```

1   **2**

🔒 **The next-gen Packt Reader** is included for free with the purchase of this book. Unlock it by scanning the QR code below or visiting `https://www.packtpub.com/unlock/9781836200918`.

The process of a diffusion transformer model goes through five main steps, as you can observe in the previous figure:

1. The visual encoder transforms datasets of images into a lower-dimensional latent space.
2. The visual encoder splits the lower-dimensional latent space into patches that are like words in a sentence.
3. The diffusion transformer associates user text input with its dictionary of patches.
4. The diffusion transformer iteratively refines noisy image representations generated to produce coherent frames.
5. The visual decoder reconstructs the refined latent representations into high-fidelity video frames that align with the user's instructions.

The video frames can then be played in a sequence. Every second of a video contains a set of frames. We will be deconstructing the AI-generated videos into frames and commenting on these frames later. But for now, we will analyze the video dataset produced by the diffusion transformer.

Analyzing the diffusion transformer model video dataset

Open the Videos_dataset_visualization.ipynb notebook on GitHub. Hopefully, you have installed the environment as described earlier in this chapter. We will move on to writing the download and display functions we need.

Video download and display functions

The three main functions each use filename (the name of the video file) as an argument. The three main functions download and display videos, and display frames in the videos.

download_video downloads one video at a time from the GitHub dataset, calling the download function defined in the *GitHub* subsection of *The environment*:

```
# downloading file from GitHub
def download_video(filename):
  # Define your variables
  directory = "Chapter10/videos"
  filename = file_name
  download(directory, filename)
```

display_video(file_name) displays the video file downloaded by first encoding in base64, a binary-to-text encoding scheme that represents binary data in ASCII string format. Then, the encoded video is displayed in HTML:

```
# Open the file in binary mode
def display_video(file_name):
    with open(file_name, 'rb') as file:
      video_data = file.read()

  # Encode the video file as base64
```

```
    video_url = b64encode(video_data).decode()

    # Create an HTML string with the embedded video
    html = f'''
    <video width="640" height="480" controls>
      <source src="data:video/mp4;base64,{video_url}" type="video/mp4">
    Your browser does not support the video tag.
    </video>
    '''

    # Display the video
    HTML(html)
    # Return the HTML object
    return HTML(html)
```

display_video_frame takes file_name, frame_number, and size (the image size to display) as arguments to display a frame in the video. The function first opens the video file and then extracts the frame number set by frame_number:

```
def display_video_frame(file_name, frame_number, size):
    # Open the video file
    cap = cv2.VideoCapture(file_name)

    # Move to the frame_number
    cap.set(cv2.CAP_PROP_POS_FRAMES, frame_number)

    # Read the frame
    success, frame = cap.read()
    if not success:
      return "Failed to grab frame"
```

The function converts the file from the BGR (blue, green, and red) to the RGB (red, green, and blue) channel, converts it to PIL, an image array (such as one handled by OpenCV), and resizes it with the size parameters:

```
    # Convert the color from BGR to RGB
    frame = cv2.cvtColor(frame, cv2.COLOR_BGR2RGB)

    # Convert to PIL image and resize
    img = Image.fromarray(frame)
    img = img.resize(size, Image.LANCZOS)  # Resize image to specified size
```

Finally, the function encodes the image in string format with base64 and displays it in HTML:

```
    # Convert the PIL image to a base64 string to embed in HTML
    buffered = BytesIO()
```

```
        img.save(buffered, format="JPEG")
        img_str = base64.b64encode(buffered.getvalue()).decode()
        # Create an HTML string with the embedded image
        html_str = f'''
        <img src="data:image/jpeg;base64,{img_str}" width="{size[0]}"
height="{size[1]}">
        '''
        # Display the image
        display(HTML(html_str))
        # Return the HTML object for further use if needed
        return HTML(html_str)
```

Once the environment is installed and the video processing functions are ready, we will display the introduction video.

Introduction video (with audio)

The following cells download and display the introduction video using the functions we created in the previous section. A video file is selected and downloaded with the download_video function:

```
# select file
print("Collecting video")
file_name="AI_Professor_Introduces_New_Course.mp4"
#file_name = "AI_Professor_Introduces_New_Course.mp4" # Enter the name of the
video file to process here
print(f"Video: {file_name}")


# Downloading video
print("Downloading video: downloading from GitHub")
download_video(file_name)
```

The output confirms the selection and download status:

```
Collecting video
Video: AI_Professor_Introduces_New_Course.mp4
Downloading video: downloading from GitHub
Downloaded 'AI_Professor_Introduces_New_Course.mp4' successfully.
```

We can choose to display only a single frame of the video as a thumbnail with the display_video_frame function by providing the file name, the frame number, and the image size to display. The program will first compute frame_count (the number of frames in the video), frame_rate (the number of frames per second), and video_duration (the duration of the video). Then, it will make sure frame_number (the frame we want to display) doesn't exceed frame_count. Finally, it displays the frame as a thumbnail:

```
print("Displaying a frame of video: ",file_name)

video_capture = cv2.VideoCapture(file_name)
frame_count = int(video_capture.get(cv2.CAP_PROP_FRAME_COUNT))
print(f'Total number of frames: {frame_count}')

frame_rate = video_capture.get(cv2.CAP_PROP_FPS)
print(f"Frame rate: {frame_rate}")

video_duration = frame_count / frame_rate
print(f"Video duration: {video_duration:.2f} seconds")

video_capture.release()
print(f'Total number of frames: {frame_count}')

frame_number=5
if frame_number > frame_count and frame_count>0:
  frame_number = 1

display_video_frame(file_name, frame_number, size=(135, 90));
```

Here, frame_number is set to 5, but you can choose another value. The output shows the information on the video and the thumbnail:

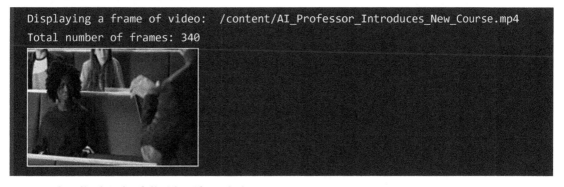

We can also display the full video if needed:

```
#print("Displaying video: ",file_name)
display_video(file_name)
```

The video will be displayed and can be played with the audio track:

Figure 10.4: AI-generated video

Let's describe and display AI-generated videos in the /videos directory of this chapter's GitHub directory. You can host this dataset in another location and scale it to the volume that meets your project's specifications. The educational video dataset of this chapter is listed in lfiles:

```
lfiles = [
    "jogging1.mp4",
    "jogging2.mp4",
    "skiing1.mp4",
    ...
    "female_player_after_scoring.mp4",
    "football1.mp4",
    "football2.mp4",
    "hockey1.mp4"
]
```

We can now move on and display any video we wish.

Displaying thumbnails and videos in the AI-generated dataset

This section is a generalization of the *Introduction video (with audio)* section. This time, instead of downloading one video, it downloads all the videos and displays the thumbnails of all the videos. You can then select a video in the list and display it.

The program first collects the video dataset:

```
for i in range(lf):
    file_name=lfiles[i]
    print("Collecting video",file_name)
    print("Downloading video",file_name)
    download_video(file_name)
```

The output shows the file names of the downloaded videos:

```
Collecting video jogging1.mp4
Downloading video jogging1.mp4
Downloaded 'jogging1.mp4' successfully.
Collecting video jogging2.mp4…
```

The program calculates the number of videos in the list:

```
lf=len(lfiles)
```

The program goes through the list and displays the information for each video and displays its thumbnail:

```
for i in range(lf):
    file_name=lfiles[i]

    video_capture.release()
    display_video_frame(file_name, frame_number=5, size=(100, 110))
```

The information on the video and its thumbnail is displayed:

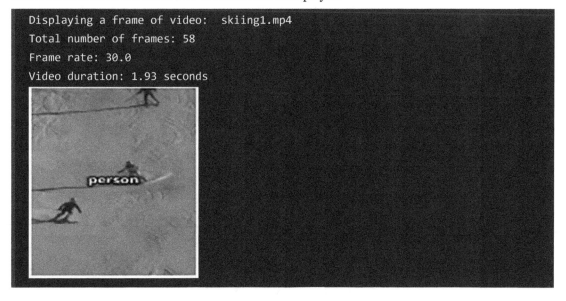

💡 **Quick tip:** Enhance your coding experience with the **AI Code Explainer** and **Quick Copy** features. Open this book in the next-gen Packt Reader. Click the **Copy** button (**1**) to quickly copy code into your coding environment, or click the **Explain** button (**2**) to get the AI assistant to explain a block of code to you.

```
                                                     Copy      Explain
function calculate(a, b) {                            1           2
  return {sum: a + b};
};
```

🔒 **The next-gen Packt Reader** is included for free with the purchase of this book. Unlock it by scanning the QR code below or visiting `https://www.packtpub.com/unlock/9781836200918`.

You can select a video in the list and display it:

```
file_name="football1.mp4" # Enter the name of the video file to process here
#print("Displaying video: ",file_name)
display_video(file_name)
```

You can click on the video and watch it:

Figure 10.5: Video of a football player

We have explored how the AI-generated videos were produced and visualized the dataset. We are now ready to build the Generator and the Commentator.

The Generator and the Commentator

The dataset of AI-generated videos is ready. We will now build the Generator and the Commentator, which processes one video at a time, making scaling seamless. An indefinite number of videos can be processed one at a time, requiring only a CPU and limited disk space. The Generator and the Commentator work together, as shown in *Figure 10.8*. These AI agents will produce raw videos from text and then split them into frames that they will comment on:

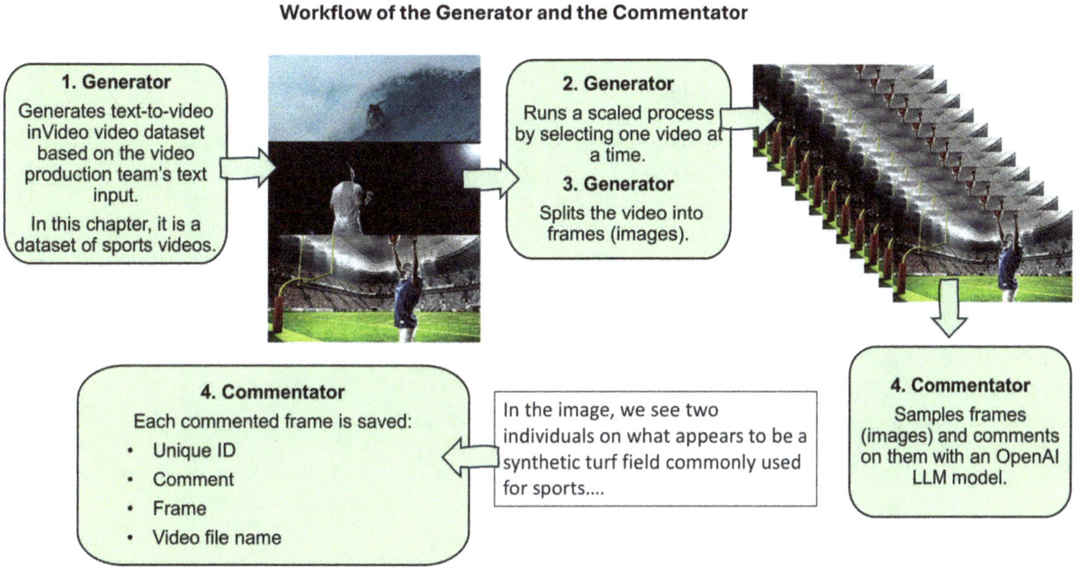

Figure 10.6: The Generator and the Commentator work together to comment on video frames

The Generator and the Commentator produce the commented frames required in four main steps that we will build in Python:

1. The **Generator** generates the text-to-video inVideo video dataset based on the video production team's text input. In this chapter, it is a dataset of sports videos.

2. The **Generator** runs a scaled process by selecting one video at a time.

3. The **Generator** splits the video into frames (images)

4. The **Commentator** samples frames (images) and comments on them with an OpenAI LLM model. Each commented frame is saved with:

    *   Unique ID
    *   Comment
    *   Frame
    *   Video file name

We will now build the Generator and the Commentator in Python, starting with the AI-generated videos. Open `Pipeline_1_The_Generator_and_the_Commentator.ipynb` in the chapter's GitHub directory. See the *The environment* section of this chapter for a description of the *Installing the environment* section of this notebook. The process of going from a video to comments on a sample of frames only takes three straightforward steps in Python:

1. Displaying the video
2. Splitting the video into frames
3. Commenting on the frames

We will define functions for each step and call them in the `Pipeline-1 Controller` section of the program. The first step is to define a function to display a video.

Step 1. Displaying the video

The download function is in the *GitHub* subsection of the *Installing the environment* section of this notebook. It will be called by the *Vector Store Administrator-Pipeline 1* in the *Administrator-Pipeline 1* section of this notebook on GitHub.

`display_video(file_name)` is the same as defined in the previous section, *The AI-generated video dataset*:

```
# Open the file in binary mode
def display_video(file_name):
    with open(file_name, 'rb') as file:
        video_data = file.read()
...

    # Return the HTML object
    return HTML(html)
```

The downloaded video will now be split into frames.

Step 2. Splitting video into frames

The `split_file(file_name)` function extracts frames from a video, as in the previous section, *The AI-generated video dataset*. However, in this case, we will expand the function to save frames as JPEG files:

```
def split_file(file_name):
    video_path = file_name
    cap = cv2.VideoCapture(video_path)

    frame_number = 0
    while cap.isOpened():
        ret, frame = cap.read()
        if not ret:
            break

        cv2.imwrite(f"frame_{frame_number}.jpg", frame)
```

```
        frame_number += 1
        print(f"Frame {frame_number} saved.")


    cap.release()
```

We have split the video into frames and saved them as JPEG images with their respective frame number, frame_number. The Generator's job finishes here and the Commentator now takes over.

Step 3. Commenting on the frames

The Generator has gone from text-to-video to splitting the video and saving the frames as JPEG frames. The Commentator now takes over to comment on the frames with three functions:

- generate_openai_comments(filename) asks the GPT-4 series vision model to analyze a frame and produce a response that contains a comment describing the frame
- generate_comment(response_data) extracts the comment from the response
- save_comment(comment, frame_number, file_name) saves the comment

We need to build the Commentator's extraction function first:

```
def generate_comment(response_data):
    """Extract relevant information from GPT-4 Vision response."""
    try:
        caption = response_data.choices[0].message.content
        return caption
    except (KeyError, AttributeError):
        print("Error extracting caption from response.")
        return "No caption available."
```

We then write a function to save the extracted comment in a CSV file that bears the same name as the video file:

```
def save_comment(comment, frame_number, file_name):
    """Save the comment to a text file formatted for seamless loading into a
pandas DataFrame."""
    # Append .csv to the provided file name to create the complete file name
    path = f"{file_name}.csv"
    # Check if the file exists to determine if we need to write headers
    write_header = not os.path.exists(path)

    with open(path, 'a', newline='') as f:
        writer = csv.writer(f, delimiter=',', quotechar='"', quoting=csv.QUOTE_
MINIMAL)
        if write_header:
            writer.writerow(['ID', 'FrameNumber', 'Comment', 'FileName'])  #
Write the header if the file is being created
```

```
        # Generate a unique UUID for each comment
        unique_id = str(uuid.uuid4())
        # Write the data
        writer.writerow([unique_id, frame_number, comment, file_name])
```

The goal is to save the comment in a format that can directly be upserted to Pinecone:

- ID: A unique string ID generated with `str(uuid.uuid4())`
- FrameNumber: The frame number of the commented JPEG
- Comment: The comment generated by the OpenAI vision model
- FileName: The name of the video file

The Commentator's main function is to generate comments with the OpenAI vision model. However, in this program's scenario, we will not save all the frames but a sample of the frames. The program first determines the number of frames to process:

```
def generate_openai_comments(filename):
  video_folder = "/content"  # Folder containing your image frames
  total_frames = len([file for file in os.listdir(video_folder) if file.
endswith('.jpg')]
```

Then, a sample frequency is set that can be modified along with a counter:

```
  nb=3      # sample frequency
  counter=0 # sample frequency counter
```

The Commentator will then go through the sampled frames and request a comment:

```
  for frame_number in range(total_frames):
      counter+=1 # sampler
      if counter==nb and counter<total_frames:
        counter=0
        print(f"Analyzing frame {frame_number}...")
        image_path = os.path.join(video_folder, f"frame_{frame_number}.jpg")
        try:
            with open(image_path, "rb") as image_file:
                image_data = image_file.read()

            response = openai.ChatCompletion.create(
                model="gpt-4-vision-preview",
```

The message is very concise: "What is happening in this image?" The message also includes the image of the frame:

```
                messages=[
                  {
                      "role": "user",
```

```
                                    "content": [
                                        {"type": "text", "text": "What is happening in
    this image?"},
                                        {
                                            "type": "image",
                                            "image_url": f"data:image/
    jpeg;base64,{base64.b64encode(image_data).decode('utf-8')}"
                                        },
                                    ],
                            }
                        ],
                        max_tokens=150,
                    )
```

Once a response is returned, the generate_comment and save_comment functions are called to extract and save the comment, respectively:

```
            comment = generate_comment(response)
            save_comment(comment, frame_number,file_name)

    except FileNotFoundError:
        print(f"Error: Frame {frame_number} not found.")
    except Exception as e:
        print(f"Unexpected error: {e}")
```

The final function we require of the Commentator is to display the comments by loading the CSV file produced in a pandas DataFrame:

```
# Read the video comments file into a pandas DataFrame
def display_comments(file_name):
    # Append .csv to the provided file name to create the complete file name
    path = f"{file_name}.csv"
    df = pd.read_csv(path)
    return df
```

The function returns the DataFrame with the comments. An administrator controls *Pipeline 1*, the Generator, and the Commentator.

Pipeline 1 controller

The controller runs jobs for the preceding three steps of the Generator and the Commentator. It begins with *Step 1*, which includes selecting a video, downloading it, and displaying it. In an automated pipeline, these functions can be separated. For example, a script would iterate through a list of videos, automatically select each one, and encapsulate the controller functions. In this case, in a pre-production and educational context, we will collect, download, and display the videos one by one:

```
session_time = time.time()  # Start timing before the request


# Step 1: Displaying the video
# select file
print("Step 1: Collecting video")
file_name = "skiing1.mp4" # Enter the name of the video file to process here
print(f"Video: {file_name}")


# Downloading video
print("Step 1:downloading from GitHub")
directory = "Chapter10/videos"
download(directory,file_name)


# Displaying video
print("Step 1:displaying video")
display_video(file_name)
```

The controller then splits the video into frames and comments on the frames of the video:

```
# Step 2.Splitting video
print("Step 2: Splitting the video into frames")
split_file(file_name)
```

The controller activates the Generator to produce comments on frames of the video:

```
# Step 3.Commenting on the video frames
print("Step 3: Commenting on the frames")
start_time = time.time()  # Start timing before the request
generate_openai_comments(file_name)
response_time = time.time() - session_time  # Measure response time
```

The response time is measured as well. The controller then adds additional outputs to display the number of frames, the comments, the content generation time, and the total controller processing times:

```
# number of frames
video_folder = "/content"  # Folder containing your image frames
total_frames = len([file for file in os.listdir(video_folder) if file.
endswith('.jpg')])
print(total_frames)
# Display comments
print("Commenting video: displaying comments")
display_comments(file_name)


total_time = time.time() - start_time  # Start timing before the request
```

```
print(f"Response Time: {response_time:.2f} seconds")  # Print response time
print(f"Total Time: {total_time:.2f} seconds")  # Print response time
```

The controller has completed its task of producing content. However, depending on your project, you can introduce dynamic RAG for some or all the videos. If you need this functionality, you can apply the process described in *Chapter 5, Boosting RAG Performance with Expert Human Feedback*, to the Commentator's outputs, including the cosine similarity quality control metrics, as we will in the *Pipeline 3: The Video Expert* section of this chapter.

The controller can also save the comments and frames.

Saving comments

To save the comments, set `save=True`. To save the frames, set `save_frames=True`. Set both values to `False` if you just want to run the program and view the outputs, but, in our case, we will set them as `True`:

```
# Ensure the file exists and double checking before saving the comments
save=True        # double checking before saving the comments
save_frames=True # double checking before saving the frames
```

The comment is saved in CSV format in `cpath` and contains the file name with the `.csv` extension and in the location of your choice. In this case, the files are saved on Google Drive (make sure the path exists):

```
# Save comments
if save==True:  # double checking before saving the comments
  # Append .csv to the provided file name to create the complete file name
  cpath = f"{file_name}.csv"
  if os.path.exists(cpath):
      # Use the Python variable 'path' correctly in the shell command
      !cp {cpath} /content/drive/MyDrive/files/comments/{cpath}
      print(f"File {cpath} copied successfully.")
  else:
      print(f"No such file: {cpath}")
```

The output confirms that a file is saved:

```
File alpinist1.mp4.csv copied successfully.
```

The frames are saved in a root name direction, for which we remove the extension with root_name = root_name + extension.strip('.'):

```python
# Save frames
import shutil
if save_frames==True:
  # Extract the root name by removing the extension
  root_name, extension = os.path.splitext(file_name)
  # This removes the period from the extension
  root_name = root_name + extension.strip('.')
  # Path where you want to copy the jpg files
  target_directory = f'/content/drive/MyDrive/files/comments/{root_name}'
  # Ensure the directory exists
  os.makedirs(target_directory, exist_ok=True)
  # Assume your jpg files are in the current directory. Modify this as needed
  source_directory = os.getcwd()  # or specify a different directory
  # List all jpg files in the source directory
  for file in os.listdir(source_directory):
      if file.endswith('.jpg'):
          shutil.copy(os.path.join(source_directory, file), target_directory)
```

The output is a directory with all the frames generated in it. We should delete the files if the controller runs in a loop over all the videos in a single session.

Deleting files

To delete the files, just set delf=True:

```python
delf=False  # double checking before deleting the files in a session
if delf==True:
  !rm -f *.mp4 # video files
  !rm -f *.jpg # frames
  !rm -f *.csv # comments
```

You can now process an unlimited number of videos one by one and scale to whatever size you wish, as long as you have disk space and a CPU!

Pipeline 2: The Vector Store Administrator

The Vector Store Administrator AI agent performs the tasks we implemented in *Chapter 6, Scaling RAG Bank Customer Data with Pinecone*. The novelty in this section relies on the fact that all the data we upsert for RAG is AI-generated. Let's open `Pipeline_2_The_Vector_Store_Administrator.ipynb` in the GitHub repository. We will build the Vector Store Administrator on top of the Generator and the Commentator AI agents in four steps, as illustrated in the following figure:

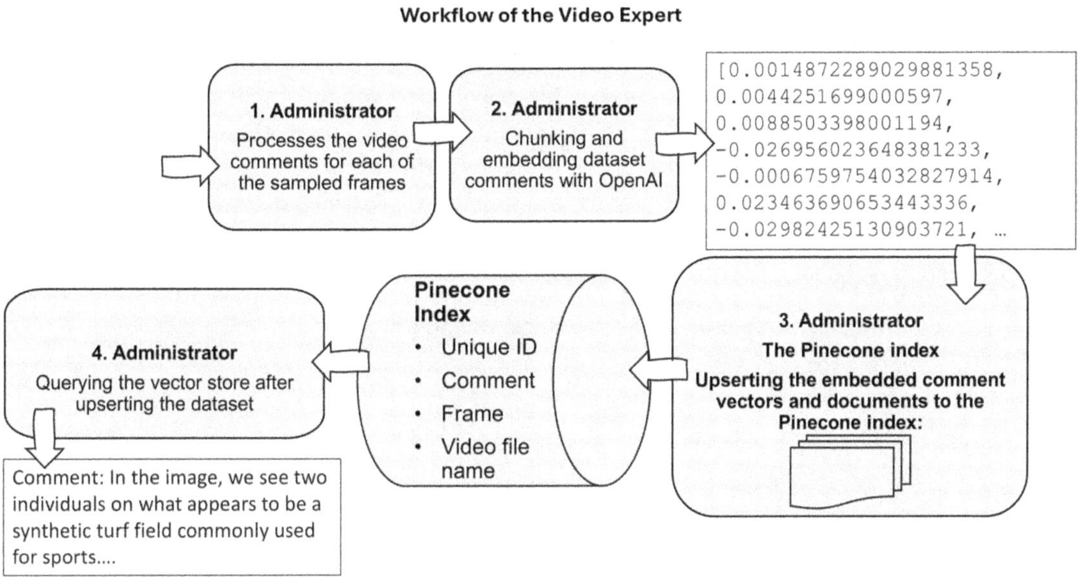

Workflow of the Video Expert

Figure 10.7: Workflow of the Vector Store Administrator from processing to querying video frame comments

1. **Processing the video comments:** The Vector Store Administrator will load and prepare the comments for chunking as in the *Pipeline 2: Scaling a Pinecone Index (vector store)* section of *Chapter 6*. Since we are processing one video at a time in a pipeline, the system deletes the files processed, which keeps disk space constant. You can enhance the functionality and scale this process indefinitely.

2. **Chunking and embedding the dataset:** The column names (`'ID'`, `'FrameNumber'`, `'Comment'`, `'FileName'`) of the dataset have already been prepared by the Commentator AI agent in *Pipeline 1*. The program chunks and embeds the dataset using the same functionality as in *Chapter 6* in the *Chunking and embedding the dataset* section.

3. **The Pinecone index:** The Pinecone Index is created, and the data is upserted as in the *Creating the Pinecone Index* and *Upserting* sections of *Chapter 6*.

4. **Querying the vector store after upserting the dataset:** This follows the same process as in *Chapter 6*. However, in this case, the retrieval is hybrid, using both the Pinecone vector store and a separate file system to store videos and video frames.

Go through *Steps 1* to *3* in the notebook to examine the Vector Store Administrator's functions. After *Step 3*, the Pinecone index is ready for hybrid querying.

Querying the Pinecone index

In the notebook on GitHub, *Step 4: Querying the Pinecone index* implements functions to find a comment that matches user input and trace it to the frame of a video. This leads to the video source and frame, which can be displayed. We can display the videos and frames from the location we wish. This hybrid approach thus involves querying the Pinecone Index to retrieve information and also retrieve media files from another location.

We saw that a vector store can contain images that are queried, as implemented in *Chapter 4, Multimodal Modular RAG for Drone Technology*. In this chapter, the video production use case videos and frame files are stored separately. In this case, it is in the GitHub repository. In production, the video and frame files can be retrieved from any storage system we need, which may or may not prove to be more cost-effective than storing data on Pinecone. The decision to store images in a vector store or a separate location will depend on the project's needs.

We begin by defining the number of top-k results we wish to process:

```
k=1 # number of results
```

We then design a rather difficult prompt:

```
query_text = "Find a basketball player that is scoring with a dunk."
```

Only a handful of frames in the whole video dataset contain an image of a basketball player jumping to score a slam dunk. Can our system find it? Let's find out.

We first embed our query to match the format of the data in the vector store:

```
import time
# Start timing before the request
start_time = time.time()


# Target vector
#query_text = "Find a basketball player."
query_embedding = get_embedding(query_text, model=embedding_model)
```

Then we run a similarity vector search between the query and the dataset:

```
# Perform the query using the embedding
query_results = index.query(vector=query_embedding, top_k=k, include_
metadata=True)  # Request metadata


# Print the query results along with metadata
print("Query Results:")
for match in query_results['matches']:
```

```
        print(f"ID: {match['id']}, Score: {match['score']}")

    # Check if metadata is available
    if 'metadata' in match:
        metadata = match['metadata']
        text = metadata.get('text', "No text metadata available.")
        frame_number = metadata.get('frame_number', "No frame number
available.")
        file_name = metadata.get('file_name', "No file name available.")
```

Finally, we display the content of the response and the response time:

```
        print(f"Text: {text}")
        print(f"Frame Number: {frame_number}")
        print(f"File Name: {file_name}")
    else:
        print("No metadata available.")

# Measure response time
response_time = time.time() - start_time
print(f"Querying response time: {response_time:.2f} seconds")  # Print response
time
```

The output contains the ID of the comment retrieved and its score:

```
Query Results:
ID: f104138b-0be8-4f4c-bf99-86d0eb34f7ee, Score: 0.866656184
```

The output also contains the comment generated by the OpenAI LLM (the Commentator agent):

```
Text: In this image, there is a person who appears to be in the process of
executing a dunk in basketball. The individual is airborne, with one arm
extended upwards towards the basketball hoop, holding a basketball in hand,
preparing to slam it through the hoop. The word "dunk" is superimposed on the
image, confirming the action taking place. The background shows clear skies
and a modern building, suggesting this might be an outdoor basketball court in
an urban setting. The player is wearing athletic wear and a pair of basketball
shoes, suitable for the sport. The dynamic posture and the context indicate an
athletic and powerful movement, typical of a basketball dunk.
```

The final output contains the frame number that was commented, the video file of the frame, and the retrieval time:

```
Frame Number: 191
File Name: basketball3.mp4
Querying response time: 0.57 seconds
```

We can display the video by downloading it based on the file name:

```python
print(file_name)
# downloading file from GitHub
directory = "Chapter10/videos"
filename = file_name
download(directory,file_name)
```

Then, use a standard Python function to display it:

```python
# Open the file in binary mode
def display_video(file_name):
  with open(file_name, 'rb') as file:
      video_data = file.read()
  # Encode the video file as base64
  video_url = b64encode(video_data).decode()
  # Create an HTML string with the embedded video
  html = f'''
  <video width="640" height="480" controls>
    <source src="data:video/mp4;base64,{video_url}" type="video/mp4">
  Your browser does not support the video tag.
  </video>
  '''

  # Display the video
  HTML(html)
  # Return the HTML object
  return HTML(html)
display_video(file_name)
```

The video containing a basketball player performing a dunk is displayed:

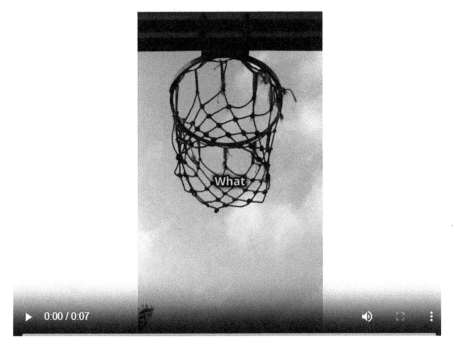

Figure 10.8: Video output

We can take this further with more precision by displaying the frame of the comment retrieved:

```
file_name_root = file_name.split('.')[0]
…
from IPython.display import Image, display
# Specify the directory and file name
directory = '/content/'  # Adjust the directory if needed
file_path = os.path.join(directory, frame)

# Check if the file exists and verify its size
if os.path.exists(file_path):
    file_size = os.path.getsize(file_path)
    print(f"File '{frame}' exists. Size: {file_size} bytes.")

    # Define a logical size value in bytes, for example, 1000 bytes
    logical_size = 1000  # You can adjust this threshold as needed

    if file_size > logical_size:
        print("The file size is greater than the logical value.")
        display(Image(filename=file_path))
```

```
    else:
        print("The file size is less than or equal to the logical value.")
else:
    print(f"File '{frame}' does not exist in the specified directory.")
```

The output shows the exact frame that corresponds to the user input:

Figure 10.9: Video frame corresponding to our input

 Only the frames of `basketball3.mp4` were saved in the GitHub repository for disk space reasons for this program. In production, all the frames you decide you need can be stored and retrieved.

The team of AI agents in this chapter worked together to generate videos (the Generator), comment on the video frames (the Commentator), upsert embedded comments in the vector store (the Vector Store Administrator), and prepare the retrieval process (the Vector Store Administrator). We also saw that the retrieval process already contained augmented input and output thanks to the OpenAI LLM (the Commentator) that generated natural language comments. The process that led to this point will definitely be applied in many domains: firefighting, medical imagery, marketing, and more.

What more can we expect from this system? The Video Expert AI agent will answer that.

Pipeline 3: The Video Expert

The role of the OpenAI GPT-4o Video Expert is to analyze the comment made by the Commentator OpenAI LLM agent, point out the cognitive dissonances (things that don't seem to fit together in the description), rewrite the comment, and provide a label. The workflow of the Video Expert, as illustrated in the following figure, also includes the code of the *Metrics calculations and display* section of *Chapter 7, Building Scalable Knowledge-Graph-Based RAG with Wikipedia API and LlamaIndex*.

The Commentator's role was only to describe what it saw. The Video Expert is there to make sure it makes sense and also label the videos so they can be classified in the dataset for further use.

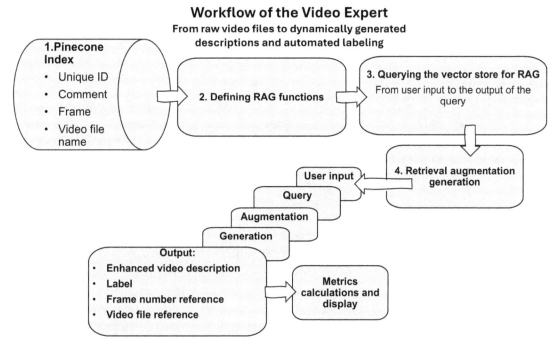

Figure 10.10: Workflow of the Video Expert for automated dynamics descriptions and labeling

1. **The Pinecone index** will connect to the Pinecone index as described in the *Pipeline 2. The Vector Store Administrator* section of this chapter. This time, we will not upsert data but connect to the vector store.

2. **Define the RAG functions** utilizing the straightforward functions we built in *Pipeline 1* and *Pipeline 2* of this chapter.

3. **Querying the vector store** is nothing but querying the Pinecone Index as described in *Pipeline 2* of this chapter.

4. **Retrieval augmented generation** finally determines the main role of Video Expert GPT-4o, which is to analyze and improve the vector store query responses. This final step will include evaluation and metric functions.

There are as many strategies as projects to implement the video production use case we explored in this chapter, but the Video Expert plays an important role. Open Pipeline_3_The_Video_Expert.ipynb on GitHub and go to the *Augmented Retrieval Generation* section in *Step 2: Defining the RAG functions*.

The function makes an OpenAI GPT-4o call, like for the Commentator in *Pipeline 1*. However, this time, the role of the LLM is quite different:

```
        "role": "system",
            "content": "You will be provided with comments of an image
frame taken from a video. Analyze the text and 1. Point out the cognitive
dissonances 2. Rewrite the comment in a logical engaging style. 3. Provide
a label for this image such as Label: basketball, football, soccer or other
label."
```

The instructions for GPT-4o are:

- `You will be provided with comments of an image frame taken from a video`: This instructs the LLM to analyze the AI-generated comments. The Commentator had to remain neutral and describe the frame as it saw it. The role of the Video Expert agent is different: it has to analyze and enhance the comment.

- `1. Point out the cognitive dissonances`: This instructs the model to find contradictions or discrepancies in the comment that can come from the way the AI-generated video was produced as well (lack of logic in the video).

- `2. Rewrite the comment in a logical engaging style`: This instructs the Video Expert agent to rewrite the comment going from a technical comment to a description.

- `3. Provide a label for this image such as Label: basketball, football, soccer or other label`: This instructs the model to provide a label for further use. On GitHub, *Step 3: Querying the Vector Store* reproduces the query and output described in *Pipeline 2* for a basketball player scoring with a dunk, with the corresponding video and frame. The output is:

```
ID=f104138b-0be8-4f4c-bf99-86d0eb34f7ee
score=0.866193652
text=In this image, there is a person who appears to be in the process
of executing a dunk in basketball. The individual is airborne, with one
arm extended upwards towards the basketball hoop, holding a basketball
in hand, preparing to slam it through the hoop. The word "dunk" is
superimposed on the image, confirming the action taking place. The
background shows clear skies and a modern building, suggesting this might
be an outdoor basketball court in an urban setting. The player is wearing
athletic wear and a pair of basketball shoes, suitable for the sport.
The dynamic posture and the context indicate an athletic and powerful
movement, typical of a basketball dunk.
frame_number=191
file_name=basketball3.mp4
```

The comment provided seems acceptable. However, let's see what GPT-4o thinks of it. The *Step 4: Retrieval Augmented Generation* section on GitHub takes the output and submits it as the user prompt to the Video Expert agent:

```
prompt=text
```

We then call the Video Expert agent to obtain its expertise:

```
response_content = get_openai_response(prompt)
print(response_content)
```

The output provides the Video Expert's insights:

```
1. Cognitive Dissonances:
    - The comment redundantly describes the action of dunking multiple times.
    - The mention of "the word 'dunk' is superimposed on the image" is
unnecessary and breaks the flow of the description.
    - The background details about clear skies and a modern building are
somewhat extraneous and could be more succinctly integrated.

2. Rewritten Comment:
    In this image, a basketball player is captured mid-air, executing a powerful
dunk. With one arm extended towards the hoop and the basketball firmly in
hand, the athlete's dynamic posture showcases the intensity of the moment. The
outdoor court, set against a backdrop of clear skies and a modern building,
highlights the urban setting. Dressed in athletic wear and basketball shoes,
the player epitomizes the energy and skill of the sport.

3. Label: Basketball
```

The response is well-structured and acceptable. The output may vary from one run to another due to the stochastic "creative" nature of Generative AI agents.

The *Evaluator* section that follows *Step 4* runs ten examples using the same process as the basketball request we just made. Each example thus contains:

- A user prompt
- The comment returned by the vector store query
- The enhanced comment made by the GPT-4o model

Each example also contains the same evaluation process as in *Chapter 7, Building Scalable Knowledge-Graph-Based RAG with Wikipedia API and LlamaIndex,* in the *Examples for metrics* section. However, in this case, the human evaluator suggests content instead of a score (0 to 1). The human content becomes the ground truth, the expected output.

Before beginning the evaluation, the program creates scores to keep track of the original response made by the query.

The human evaluator rewrites the output provided by the Video Expert:

```
# Human feedback flashcard comment
text1 = "This image shows soccer players on a field dribbling and passing the
ball."
```

The content rewritten by the Video Expert is extracted from the response:

```
# Extract rewritten comment
text2 = extract_rewritten_comment(response_content)
```

The human comment (ground truth, the reference output) and the LLM comments are displayed:

```
print(f"Human Feedback Comment: {text1}")
print(f"Rewritten Comment: {text2}")
```

Then, the cosine similarity score between the human and LLM comments is calculated and appended to scores:

```
similarity_score3=calculate_cosine_similarity_with_embeddings(text1, text2)
print(f"Cosine Similarity Score with sentence transformer: {similarity_
score3:.3f}")
scores.append(similarity_score3)
```

The original score provided with the query is appended to the query's retrieval score, rscores:

```
rscores.append(score)
```

The output displays the human feedback, the comment rewritten by GPT-4o (the Video Expert), and the similarity score:

```
Human Feedback Comment: This image shows soccer players on a field dribbling
and passing the ball.
Rewritten Comment: "A group of people are engaged in a casual game of soccer
on a grassy field. One player is dribbling the ball while others are either
defending or waiting for a pass. They are dressed in athletic attire,
indicating this is a recreational game among friends or acquaintances.
Interestingly, there is a superimposed text 'female' that seems unrelated to
the activity shown in the image."
Cosine Similarity Score with sentence transformer: 0.621
```

This program contains ten examples, but we can enter a corpus of as many examples as we wish to evaluate the system. The evaluation of each example applies the same choice of metrics as in *Chapter 7*. After the examples have been evaluated, the *Metrics calculations and display* section in the program also runs the metric calculations defined in the section of the same name in *Chapter 7*.

We can use all the metrics to analyze the performance of the system. The time measurements throughout the program also provide insights. The first metric, accuracy, is a good metric to start with. In this case, it shows that there is room for progress:

```
Mean: 0.65
```

Some requests and responses were challenging and required further work to improve the system:

•   Checking the quality of the videos and their content

- Checking the comments and possibly modifying them with human feedback, as we did in *Chapter 5, Boosting RAG Performance with Expert Human Feedback*

- Fine-tuning a model with images and text as we did in *Chapter 9, Empowering AI Models: Fine-Tuning RAG Data and Human Feedback*

- Designing any other constructive idea that the video production team comes up with

We can see that RAG-driven Generative AI systems in production are very effective. However, the road from design to production requires hard human effort! Though AI technology has made tremendous progress, it still requires humans to design, develop, and implement it in production.

Summary

In this chapter, we explored the hybrid era of human and AI agents, focusing on the creation of a streamlined process for generating, commenting, and labeling videos. By integrating cutting-edge Generative AI models, we demonstrated how to build an automated pipeline that transforms raw video inputs into structured, informative, and accessible video content.

Our journey began with the **Generator** agent in *Pipeline 1: The Generator and the Commentator*, which was tasked with creating video content from textual ideas. We can see that video generation processes will continue to expand through seamless integration ideation and descriptive augmentation generative agents. In *Pipeline 2: The Vector Store Administrator*, we focused on organizing and embedding the generated comments and metadata into a searchable vector store. In this pipeline, we highlighted the optimization process of building a scalable video content library with minimal machine resources using only a CPU and no GPU. Finally, in *Pipeline 3: The Video Expert*, we introduced the Expert AI agent, a video specialist designed to enhance and label the video content based on user inputs. We also implemented evaluation methods and metric calculations.

By the end of this chapter, we had constructed a comprehensive, automated RAG-driven Generative AI system capable of generating, commenting on, and labeling videos with minimal human intervention. This journey demonstrated the power and potential of combining multiple AI agents and models to create an efficient pipeline for video content creation.

The techniques and tools we explored can revolutionize various industries by automating repetitive tasks, enhancing content quality, and making information retrieval more efficient. This chapter not only provided a detailed technical roadmap but also underscored the transformative impact of AI in modern content creation and management. You are now all set to implement RAG-driven Generative AI in real-life projects.

Questions

Answer the following questions with yes or no:

1. Can AI now automatically comment and label videos?
2. Does video processing involve splitting the video into frames?
3. Can the programs in this chapter create a 200-minute movie?
4. Do the programs in this chapter require a GPU?

5. Are the embedded vectors of the video content stored on disk?

6. Do the scripts involve querying a database for retrieving data?

7. Is there functionality for displaying images in the scripts?

8. Is it useful to have functions that specifically check file existence and size in any of the scripts?

9. Is there a focus on multimodal data in these scripts?

10. Do any of the scripts mention applications of AI in real-world scenarios?

References

- Sora video generation model information and access:

    - **Sora | OpenAI**: https://ai.invideo.io/

    - https://openai.com/index/video-generation-models-as-world-simulators/

- *Sora: A Review on Background, Technology, Limitations, and Opportunities of Large Vision Models* by Yixin Liu, Kai Zhang, Yuan Li, et al.: https://arxiv.org/pdf/2402.17177

Further reading

- OpenAI, ChatGPT: https://openai.com/chatgpt/

- OpenAI, Research: https://openai.com/research/

- Pinecone: https://docs.pinecone.io/home

Unlock this book's exclusive benefits now

This book comes with additional benefits designed to elevate your learning experience.

Note: Have your purchase invoice ready before you begin. https://www.packtpub.com/unlock/9781836200918

Appendix

The appendix here provides answers to all questions added at the end of each chapter. Double-check your answers to verify that you have conceptually understood the key concepts.

Chapter 1, Why Retrieval Augmented Generation?

1. Is RAG designed to improve the accuracy of generative AI models?

    Yes, RAG retrieves relevant data to enhance generative AI outputs.

2. Does a naïve RAG configuration rely on complex data embedding?

    No, naïve RAG uses basic keyword searches without advanced embeddings.

3. Is fine-tuning always a better option than using RAG?

    No, RAG is better for handling dynamic, real-time data.

4. Does RAG retrieve data from external sources in real time to enhance responses?

    Yes, RAG pulls data from external sources during query processing.

5. Can RAG be applied only to text-based data?

    No, RAG works with text, images, and audio data as well.

6. Is the retrieval process in RAG triggered by a user or automated input?

    The retrieval process in RAG is typically triggered by a query, which can come from a user or an automated system.

7. Are cosine similarity and TF-IDF both metrics used in advanced RAG configurations?

    Yes, both are used to assess the relevance between queries and documents.

8. Does the RAG ecosystem include only data collection and generation components?

    No, it also includes storage, retrieval, evaluation, and training.

9. Can advanced RAG configurations process multimodal data such as images and audio?

    Yes, advanced RAG supports processing structured and unstructured multimodal data.

10. Is human feedback irrelevant in evaluating RAG systems?

    No, human feedback is crucial for improving RAG system accuracy and relevance.

Chapter 2, RAG Embedding Vector Stores with Deep Lake and OpenAI

1. Do embeddings convert text into high-dimensional vectors for faster retrieval in RAG?

   Yes, embeddings create vectors that capture the semantic meaning of text.

2. Are keyword searches more effective than embeddings in retrieving detailed semantic content?

   No, embeddings are more context-aware than rigid keyword searches.

3. Is it recommended to separate RAG pipelines into independent components?

   Yes, this allows parallel development and easier maintenance.

4. Does the RAG pipeline consist of only two main components?

   No, the pipeline consists of three components – data collection, embedding, and generation.

5. Can Activeloop Deep Lake handle both embedding and vector storage?

   Yes, it stores embeddings efficiently for quick retrieval.

6. Is the text-embedding-3-small model from OpenAI used to generate embeddings in this chapter?

   Yes, this model is chosen for its balance between detail and computational efficiency.

7. Are data embeddings visible and directly traceable in an RAG-driven system?

   Yes, unlike parametric models, embeddings in RAG are traceable to the source.

8. Can a RAG pipeline run smoothly without splitting into separate components?

   Splitting an RAG pipeline into components improves specialization, scalability, and security, which helps a system run smoothly. Simpler RAG systems may still function effectively without explicit component separation, although it may not be the optimal setup.

9. Is chunking large texts into smaller parts necessary for embedding and storage?

   Yes, chunking helps optimize embedding and improves the efficiency of queries.

10. Are cosine similarity metrics used to evaluate the relevance of retrieved information?

    Yes, cosine similarity helps measure how closely retrieved data matches the query.

Chapter 3, Building Index-Based RAG with LlamaIndex, Deep Lake, and OpenAI

1. Do indexes increase precision and speed in retrieval-augmented generative AI?

   Yes, indexes make retrieval faster and more accurate.

2. Can indexes offer traceability for RAG outputs?

   Yes, indexes allow tracing back to the exact data source.

3. Is index-based search slower than vector-based search for large datasets?

   No, index-based search is faster and optimized for large datasets.

4. Does LlamaIndex integrate seamlessly with Deep Lake and OpenAI?

   Yes, LlamaIndex, Deep Lake, and OpenAI work well together.

5. Are tree, list, vector, and keyword indexes the only types of indexes?

   No, these are common, but other types exist as well.

6. Does the keyword index rely on semantic understanding to retrieve data?

   No, it retrieves based on keywords, not semantics.

7. Is LlamaIndex capable of automatically handling chunking and embedding?

   Yes, LlamaIndex automates these processes for easier data management.

8. Are metadata enhancements crucial for ensuring the traceability of RAG-generated outputs?

   Yes, metadata helps trace back to the source of the generated content.

9. Can real-time updates easily be applied to an index-based search system?

   Indexes often require re-indexing for updates. However, some modern indexing systems have been designed to handle real-time or near-real-time updates more efficiently.

10. Is cosine similarity a metric used in this chapter to evaluate query accuracy?

    Yes, cosine similarity helps assess the relevance of query results.

Chapter 4, Multimodal Modular RAG for Drone Technology

1. Does multimodal modular RAG handle different types of data, such as text and images?

   Yes, it processes multiple data types such as text and images.

2. Are drones used solely for agricultural monitoring and aerial photography?

   No, drones are also used for rescue, traffic, and infrastructure inspections.

3. Is the Deep Lake VisDrone dataset used in this chapter for textual data only?

   No, it contains labeled drone images, not just text.

4. Can bounding boxes be added to drone images to identify objects such as trucks and pedestrians?

   Yes, bounding boxes are used to mark objects within images.

5. Does the modular system retrieve both text and image data for query responses?

   Yes, it retrieves and generates responses from both textual and image datasets.

6. Is building a vector index necessary for querying the multimodal VisDrone dataset?

    Yes, a vector index is created for efficient multimodal data retrieval.

7. Are the retrieved images processed without adding any labels or bounding boxes?

    No, images are processed with labels and bounding boxes.

8. Is the multimodal modular RAG performance metric based only on textual responses?

    No, it also evaluates the accuracy of image analysis.

9. Can a multimodal system such as the one described in this chapter handle only drone-related data?

    No, it can be adapted for other industries and domains.

10. Is evaluating images as easy as evaluating text in multimodal RAG?

    No, image evaluation is more complex and requires specialized metrics.

Chapter 5, Boosting RAG Performance with Expert Human Feedback

1. Is human feedback essential in improving RAG-driven generative AI systems?

    Yes, human feedback directly enhances the quality of AI responses.

2. Can the core data in a generative AI model be changed without retraining the model?

    No, the model's core data is fixed unless it is retrained.

3. Does Adaptive RAG involve real-time human feedback loops to improve retrieval?

    Yes, Adaptive RAG uses human feedback to refine retrieval results.

4. Is the primary focus of Adaptive RAG to replace all human input with automated responses?

    No, it aims to blend automation with human feedback.

5. Can human feedback in Adaptive RAG trigger changes in the retrieved documents?

    Yes, feedback can prompt updates to retrieved documents for better responses.

6. Does Company C use Adaptive RAG solely for customer support issues?

    No, it's also used for explaining AI concepts to employees.

7. Is human feedback used only when the AI responses have high user ratings?

    No, feedback is often used when responses are rated poorly.

8. Does the program in this chapter provide only text-based retrieval outputs?

    No, it uses both text and expert feedback for responses.

9. Is the Hybrid Adaptive RAG system static, meaning it cannot adjust based on feedback?

   No, it dynamically adjusts to feedback and rankings.

10. Are user rankings completely ignored in determining the relevance of AI responses?

   No, user rankings directly influence the adjustments made to a system.

Chapter 6, Scaling RAG Bank Customer Data with Pinecone

1. Does using a Kaggle dataset typically involve downloading and processing real-world data for analysis?

   Yes, Kaggle datasets are used for practical, real-world data analysis and modeling.

2. Is Pinecone capable of efficiently managing large-scale vector storage for AI applications?

   Yes, Pinecone is designed for large-scale vector storage, making it suitable for complex AI tasks.

3. Can k-means clustering help validate relationships between features such as customer complaints and churn?

   Yes, k-means clustering is useful for identifying and validating patterns in datasets.

4. Does leveraging over a million vectors in a database hinder the ability to personalize customer interactions?

   No, handling large volumes of vectors allows for more personalized and targeted customer interactions.

5. Is the primary objective of using generative AI in business applications to automate and improve decision-making processes?

   Yes, generative AI aims to automate and refine decision-making in various business applications.

6. Are lightweight development environments advantageous for rapid prototyping and application development?

   Yes, they streamline development processes, making it easier and faster to test and deploy applications.

7. Can Pinecone's architecture automatically scale to accommodate increasing data loads without manual intervention?

   Yes, Pinecone's serverless architecture supports automatic scaling to handle larger data volumes efficiently.

8. Is generative AI typically employed to create dynamic content and recommendations based on user data?

   Yes, generative AI is often used to generate customized content and recommendations dynamically.

9. Does the integration of AI technologies such as Pinecone and OpenAI require significant manual configuration and maintenance?

   No, these technologies are designed to minimize manual efforts in configuration and maintenance through automation.

10. Are projects that use vector databases and AI expected to effectively handle complex queries and large datasets?

    Yes, vector databases combined with AI are particularly well-suited for complex queries and managing large datasets.

Chapter 7, Building Scalable Knowledge-Graph-based RAG with Wikipedia API and LlamaIndex

1. Does the chapter focus on building a scalable knowledge graph-based RAG system using the Wikipedia API and LlamaIndex?

   Yes, it details creating a knowledge graph-based RAG system using these tools.

2. Is the primary use case discussed in the chapter related to healthcare data management?

   No, the primary use case discussed is related to marketing and other domains.

3. Does Pipeline 1 involve collecting and preparing documents from Wikipedia using an API?

   Yes, Pipeline 1 automates document collection and preparation using the Wikipedia API.

4. Is Deep Lake used to create a relational database in Pipeline 2?

   No, Deep Lake is used to create and populate a vector store, not a relational database.

5. Does Pipeline 3 utilize LlamaIndex to build a knowledge graph index?

   Yes, Pipeline 3 uses LlamaIndex to build a knowledge graph index automatically.

6. Is the system designed to only handle a single specific topic, such as marketing, without flexibility?

   No, the system is flexible and can handle various topics beyond marketing.

7. Does the chapter describe how to retrieve URLs and metadata from Wikipedia pages?

   Yes, it explains the process of retrieving URLs and metadata using the Wikipedia API.

8. Is a GPU required to run the pipelines described in the chapter?

   No, the pipelines are designed to run efficiently using only a CPU.

9. Does the knowledge graph index visually map out relationships between pieces of data?

   Yes, the knowledge graph index visually displays semantic relationships in the data.

10. Is human intervention required at every step to query the knowledge graph index?

    No, querying the knowledge graph index is automated, with minimal human intervention needed.

Chapter 8, Dynamic RAG with Chroma and Hugging Face Llama

1. Does the script ensure that the Hugging Face API token is never hardcoded directly into the notebook for security reasons?

   Yes, the script provides methods to either use Google Drive or manual input for API token handling, thus avoiding hardcoding.

2. In the chapter's program, is the accelerate library used to facilitate the deployment of machine learning models on cloud-based platforms?

   No, the accelerate library is used to run models on local resources such as multiple GPUs, TPUs, and CPUs, not specifically cloud platforms.

3. Is user authentication, apart from the API token, required to access the Chroma database in this script?

   No, the script does not detail additional authentication mechanisms beyond using an API token to access Chroma.

4. Does the notebook use Chroma for temporary storage of vectors during the dynamic retrieval process?

   Yes, the script employs Chroma for storing vectors temporarily to enhance the efficiency of data retrieval.

5. Is the notebook configured to use real-time acceleration of queries through GPU optimization?

   Yes, the accelerate library is used to ensure that the notebook can leverage GPU resources for optimizing queries, which is particularly useful in dynamic retrieval settings.

6. Can this notebook's session time measurements help in optimizing the dynamic RAG process?

   Yes, by measuring session time, the notebook provides insights that can be used to optimize the dynamic RAG process, ensuring efficient runtime performance.

7. Does the script demonstrate Chroma's capability to integrate with machine learning models for enhanced retrieval performance?

   Yes, the integration of Chroma with the Llama model in this script highlights its capability to enhance retrieval performance by using advanced machine learning techniques.

8. Does the script include functionality to adjust the parameters of the Chroma database based on session performance metrics?

Yes, the notebook potentially allows adjustments to be made based on performance metrics, such as session time, which can influence how the notebook is built and adjust the process, depending on the project.

Chapter 9, Empowering AI Models: Fine-Tuning RAG Data and Human Feedback

1. Do all organizations need to manage large volumes of RAG data?

No, many corporations only need small data volumes.

2. Is the GPT-4-o-mini model described as insufficient for fine-tuning tasks?

No, GPT-4o-mini is described as cost-effective for fine-tuning tasks.

3. Can pretrained models update their knowledge base after the cutoff date without retrieval systems?

No, models are static and rely on retrieval for new information.

4. Is it the case that static data never changes and thus never requires updates?

No, Only that it remains stable for a long time, not forever.

5. Is downloading data from Hugging Face the only source for preparing datasets?

Yes, Hugging Face is specifically mentioned as the data source.

6. Are all RAG data eventually embedded into the trained model's parameters?

No, non-parametric data remains external.

7. Does the chapter recommend using only new data for fine-tuning AI models?

No, it suggests fine-tuning with relevant, often stable data.

8. Is the OpenAI metric interface used to adjust the learning rate during model training?

No, it monitors performance and costs after training.

9. Can the fine-tuning process be effectively monitored using the OpenAI dashboard?

Yes, the dashboard provides real-time updates on fine-tuning jobs.

10. Is human feedback deemed unnecessary in the preparation of hard science datasets such as SciQ?

No, human feedback is crucial for data accuracy and relevance.

Chapter 10, RAG for Video Stock Production with Pinecone and OpenAI

1. Can AI now automatically comment and label videos?

   Yes, we now create video stocks automatically to a certain extent.

2. Does video processing involve splitting a video into frames?

   Yes, we can split a video into frames before analyzing the frames.

3. Can the programs in this chapter create a 200-minute movie?

   No, for the moment, this cannot be done directly. We would have to create many videos and then stitch them together with a video editor.

4. Do the programs in this chapter require a GPU?

   No, only a CPU is required, which is cost-effective because the processing times are reasonable, and the programs mostly rely on API calls.

5. Are the embedded vectors of the video content stored on disk?

   No, the embedded vectors are upserted in a Pinecone vector database.

6. Do the scripts involve querying a database for retrieving data?

   Yes, the scripts query the Pinecone vector database for data retrieval.

7. Is there functionality for displaying images in the scripts?

   Yes, the programs include code to display images after downloading them.

8. Is it useful to have functions specifically checking file existence and size in any of the scripts?

   Yes, this avoids trying to display files that don't exist or that are empty.

9. Is there a focus on multimodal data in these scripts?

   Yes, all scripts focus on handling and processing multimodal data (text, image, and video).

10. Do any of the scripts mention applications of AI in real-world scenarios?

    Yes, these scripts deal with multimodal data retrieval and processing, which makes them applicable in AI-driven content management, search, and retrieval systems.

Unlock this book's exclusive benefits now

This book comes with additional benefits designed to elevate your learning experience.

Note: Have your purchase invoice ready before you begin. `https://www.packtpub.com/unlock/9781836200918`

packt.com

Subscribe to our online digital library for full access to over 7,000 books and videos, as well as industry leading tools to help you plan your personal development and advance your career. For more information, please visit our website.

Why subscribe?

- Spend less time learning and more time coding with practical eBooks and Videos from over 4,000 industry professionals
- Improve your learning with Skill Plans built especially for you
- Get a free eBook or video every month
- Fully searchable for easy access to vital information
- Copy and paste, print, and bookmark content

At www.packt.com, you can also read a collection of free technical articles, sign up for a range of free newsletters, and receive exclusive discounts and offers on Packt books and eBooks.

Other Books You May Enjoy

If you enjoyed this book, you may be interested in these other books by Packt:

Transformers for Natural Language Processing and Computer Vision - Third Edition

Denis Rothman

ISBN: 9781805128724

- Breakdown and understand the architectures of the Original Transformer, BERT, GPT models, T5, PaLM, ViT, CLIP, and DALL-E
- Fine-tune BERT, GPT, and PaLM 2 models
- Learn about different tokenizers and the best practices for preprocessing language data
- Pretrain a RoBERTa model from scratch
- Implement retrieval augmented generation and rules bases to mitigate hallucinations
- Visualize transformer model activity for deeper insights using BertViz, LIME, and SHAP
- Go in-depth into vision transformers with CLIP, DALL-E 2, DALL-E 3, and GPT-4V

Generative AI Application Integration Patterns

Juan Pablo Bustos, Luis Lopez Soria

ISBN: 9781835887608

- Concepts of GenAI: pre-training, fine-tuning, prompt engineering, and RAG
- Framework for integrating AI: entry points, prompt pre-processing, inference, post-processing, and presentation
- Patterns for batch and real-time integration
- Code samples for metadata extraction, summarization, intent classification, question-answering with RAG, and more
- Ethical use: bias mitigation, data privacy, and monitoring
- Deployment and hosting options for GenAI models

Packt is searching for authors like you

If you're interested in becoming an author for Packt, please visit `authors.packtpub.com` and apply today. We have worked with thousands of developers and tech professionals, just like you, to help them share their insight with the global tech community. You can make a general application, apply for a specific hot topic that we are recruiting an author for, or submit your own idea.

Share your thoughts

Now you've finished *RAG-Driven Generative AI*, we'd love to hear your thoughts! Scan the QR code below to go straight to the Amazon review page for this book and share your feedback or leave a review on the site that you purchased it from.

https://packt.link/r/1836200919

Your review is important to us and the tech community and will help us make sure we're delivering excellent quality content.

Index

www.ingramcontent.com/pod-product-compliance
Lightning Source LLC
LaVergne TN
LVHW082125070326
832902LV00041B/2547